SUSTAINABLE MARKETING OF TRANSFORMATIVE HERITAGE TOURISM

This book proposes that sustainable marketing should be founded on a higher level of consciousness, governed by the Eastern ethical principles of dharma and karma. It suggests a socially responsible, integrated marketing strategy to boost transformational heritage tourism.

The book puts forward a framework to promote and inspire transformative heritage tourism from a sustainable marketing perspective. Drawing on examples from different parts of the world such as Thailand, China, the USA, India, Australia, the United Kingdom, Spain and Hong Kong, it takes a unique approach by integrating elements associated with distributive justice, procedural justice, morally motivated existentialist authentic experiences and the well-being of visited and visiting environments. The book pivots on the planning and marketing of heritage of communities such as local, descendent and indigenous across eight broad themes: (1) promotion and sustainable branding of heritage tourism; (2) empowerment of indigenous communities; (3) authenticity and conservation of heritage; (4) safeguarding of art, culture and cultural landscapes; (5) economic viability for the host communities; (6) interpretation and resolution of dissonant heritage representations; (7) stimulating audience engagement and co-created mindful spaces and (8) facilitating moral consciousness and stimulating lasting inner transformation in guests and hosts. Scholars can replicate and/or test the proposed guided sustainable marketing model, referred to as the transformative heritage tourism pathway, which is adapted from a) the needs, activities and well-being sequential path of transformation and, b) the socially responsible sustainable marketing doctrine guided by the principles of 'dharma' and 'karma.' This book is unique as it stresses eudaimonia as the ultimate goal of well-being and argues that its pursuit can steer the sustained transformation route toward a higher sense of consciousness and responsible production and consumption of heritage resources.

In summary, this book contributes toward advancing the dialogue on sustainable marketing and transformative heritage tourism. It will appeal to researchers, undergraduates and practitioners interested in sustainable marketing, transformative heritage tourism and social, ecological, cultural and economic welfare of visited and visiting destinations.

Deepak Chhabra is Associate Professor in the School of Community Resources and Development (at Arizona State University, Phoenix, USA) who specializes in socio-economic impacts of tourism and sustainable management and marketing of culture and heritage. She also holds the position of Global Futures Scientist at the Julie Ann Wrigley Global Futures Laboratory, Arizona State University. Her research specifically focuses on transformative paradigms, authenticity and authentication of heritage, and designing smart and sustainable marketing strategies in hospitality and tourism that can foster social, cultural and economic equity/capital in local, regional and global communities. She serves as an editorial board member for several leading journals in her field. She has authored more than sixty articles in refereed journals. She has also authored and edited several books. Her first book (on sustainable marketing of cultural and heritage tourism) is recognized throughout the world by academicians as well as practitioners. Her sustainable marketing model, from this book, has been used as a blueprint by several cultural and heritage institutions and government agencies across the globe. She recently co-edited a book titled *Sustainable Development and Resilience of Tourism*.

SUSTAINABLE MARKETING OF TRANSFORMATIVE HERITAGE TOURISM

Edited by Deepak Chhabra

Routledge
Taylor & Francis Group
LONDON AND NEW YORK

Designed cover image: Deepak Chhabra

First published 2025
by Routledge
4 Park Square, Milton Park, Abingdon, Oxon OX14 4RN

and by Routledge
605 Third Avenue, New York, NY 10158

Routledge is an imprint of the Taylor & Francis Group, an informa business

British Library Cataloguing-in-Publication Data
A catalogue record for this book is available from the British Library

ISBN: 9781032386676 (hbk)
ISBN: 9781032386683 (pbk)
ISBN: 9781003346197 (ebk)

DOI: 10.4324/9781003346197

Typeset in Sabon
by KnowledgeWorks Global Ltd.

CONTENTS

FIGURES

TABLES

CONTRIBUTORS

Jirawan Chaikor owns Ahpa Tea, Thailand. She operates a small hotel and community-based tea tourism benefiting her Lahu hill tribe village of Doi Punmeun located in northern Chiang Mai Province. She has received recognition and certification for sustainable community-based tourism and is a frequent speaker on this topic.

Deepak Chhabra is Associate Professor in the School of Community Resources and Development (at Arizona State University, Phoenix, USA) who specializes in socio-economic impacts of tourism and sustainable management and marketing of culture and heritage. She also holds the position of Global Futures Scientist at the Julie Ann Wrigley Global Futures Laboratory, Arizona State University. Her research specifically focuses on transformative paradigms, authenticity and authentication of heritage, and designing smart and sustainable marketing strategies in hospitality and tourism that can foster social, cultural, and economic equity/capital in local, regional, and global communities. She serves as an editorial board member for several leading journals in her field. She has authored more than sixty articles in refereed journals. She has also authored and edited several books. Her first book (on sustainable marketing of cultural and heritage tourism) is recognized, throughout the world, by academicians as well as the practitioners. Her sustainable marketing model, from this book, has been used as a blueprint by several cultural and heritage institutions and government agencies across the globe. She recently co-edited a book titled *Sustainable Development and Resilience of Tourism*.

Cemile Ece is a PhD candidate in the Department of Tourism and Hotel Management and project expert. She currently researches eco-tourism, sustainability, destination management, destination branding, and consumer behaviour. She has experiences in EU+ projects and National-Funded Projects.

Michael Fagence is first and foremost a geographer. In recent years, his research has focused on applying processes of 'thinking geographically' together with investigative and interpretive processes from history, literary studies and semiotics to 'tease out' opportunities for tourism development through the lens of heritage-based tourism. His most recent appointment was as an Honorary Research Fellow at the University of Queensland.

S.R. Gnanasambandan holds a post-graduate degree in Tourism and Travel Management from Amity University, Noida, and undergraduate degree in Tourism and Travel Management from Christ University, Bangalore. He worked in WM Travel Solutions Pvt. Ltd. in the MICE Department as Executive Operations. He is very much passionate about travelling and exploring different destinations along with experiencing different cultures and cuisines.

Linsi He is Postdoctoral Researcher of School of Ethnology and Sociology at Yunnan University. She is a PhD graduate from the School of Community Resources and Development at Arizona State University. Her research interests include souvenirs, authenticity, cultural tourism, place and representation.

Lee Jolliffe is a Research Advisor, Tourism Hospitality and Research Events Group, School of Management, Mae Fah Luang University, Thailand. She has written extensively on tea tourism and was the lead editor for *The Routledge Handbook of Tea Tourism* published in 2022.

Alka Maheshwari has over 32 years of experience in academics. She is working as an Associate Professor in Amity Institute of Travel and Tourism. She has done her Master's degree in English Literature and Tourism and her PhD is in Community-Based Tourism. Dr. Alka was fortunate to visit Lancashire University, Blackburn, as a team member for UKIERI. Her broad research areas of interest include Community-Based Tourism, Tourism Marketing, Sustainable Tourism, Tourism Policy and Planning, Tourism Products and Services. She has publications in various national and international journals.

Pairach Piboonrungroj is Associate Dean, Faculty of Economics, and Director of the MICE Excellence Centre and Supply Chain Research Centre at Chiang Mai University, Thailand. His supply chain research includes a focus on coffee and tea through community-based projects in Chiang Rai Province.

Rijuta Sawhney is working as a Graduate Travel and Tourism Teacher at Bexhill College. She has done her Masters in International Tourism and Aviation Management from Sheffield Hallam University and Bachelors in Tourism Administration from Amity University, Noida. She has her research interests in Technology in Travel and Tourism and Cultural Tourism.

Dallen Timothy is Professor of Community Resources and Development, and a Senior Sustainability Scientist at Arizona State University. He is also a Visiting Professor in the Erasmus European Master of Tourism Management based at the University of Girona, Spain, and Research Associate at the University of Johannesburg, South Africa.

Alba Viana-Lora holds a PhD in Tourism and Leisure from the Rovira I Virgili University. She is a member of the SEJ-588 research group at the University of Cordoba. Her lines of research focus on the social impact of research and knowledge transfer in tourism, as well as on sustainability and other social issues in the tourism industry.

Philip Xie is Professor of Tourism, Hospitality and Event Management at Bowling Green State University, Ohio, USA. He undertakes research, writes and consults in tourism planning. He has authored three books: *The Morphology of Tourism* (2019), *Industrial Heritage Tourism* (2015) and *Authenticating Ethnic Tourism* (2010). In addition, he has co-edited a book entitled *Colonialism, Tourism and Place* (2020). He is one of the leading academic experts in the field of tourism planning, who received several major grants from State and Federal governments in the US.

Orhan Can Yılmazdoğan is Assistant Professor at Eskişehir Osmangazi University, Tourism Faculty, where he teaches courses on Marketing, Digital Marketing and Cultural Tourism. His research focuses on Marketing, Social Media and Consumer Behavior. Orhan Can YILMAZDOĞAN has scientific articles and book chapters in the field of Tourism. He has experiences on scientific researches and project management of more than ten years. He is responsible for developing the content of educational courses.

ACKNOWLEDGMENTS

This book grew out of my numerous experiences of working on sustainable marketing projects and my years of teaching tourism marketing. My interest piqued after I penned down my first book on sustainable marketing of heritage tourism in 2010. Scholarship has evolved since 2010 and emphasis on sustainable marketing continues to grow particularly following the worldwide devastation caused by the pandemic and numerous closures/setbacks experienced by the public sector and many tourism enterprises, both for-profit and non-profit, across the globe. Today, the whole world recognizes the significance of embracing a transformative path to promote and resolve sustainable issues associated with niche forms of tourism. Heritage tourism is one form of niche tourism that holds tremendous potential to contribute to the economic, social, cultural, ecological and spiritual well-being of all stakeholders, particularly the host societies, in numerous ways. There is a call for all stakeholders of the heritage tourism system to collaborate mindfully and transformatively so that they can become resilient to counter the devastating effects of unforeseen crises of global magnitude. My personal interest in transformative heritage tourism and well-being of host and guest societies also offers impetus for this book.

I am enormously grateful to the scholars who contributed to this book. I also acknowledge the contributions of many other people and heritage agencies and institutions across the globe. I also thank my colleagues and administrative staff at the School of Community Resources and Development for their insights and assistance. Bulk of this research was supported by college and school funds. Furthermore, a number of experts from heritage institutions enriched my fieldwork and I am thankful to them.

Finally, I thank my family and dogs, Ebony and Sania. This book is dedicated to my late father, Dr. G.S. Chhabra and my late mother, Sharanjit Walia. They are no more with me but they continue to inspire me through the wisdom they imparted and the fond memories I will always have of them.

PREFACE

From the time I penned down my first book on *Sustainable Marketing of Heritage Tourism* in 2010, I have kept track of ongoing deliberations in the field of sustainable marketing from a cultural/heritage tourism standpoint. Surprisingly, literature is still sparse on this subject, given that a sustainable marketing agenda can tremendously assist in balancing preservation ethics and tourism demand for heritage sites and numerous other cultural heritage attractions. Management of cultural/heritage tourism sites still requires guidance with their sustainable marketing plans. Such plans call for a long-term vision and both time and financial investment as they can impact a broad spectrum of stakeholders, particularly the local residents.

In tracing the path of sustainable marketing in tourism literature, it is noted that in 1999, Fuller proposed one of the first sustainable marketing management paradigms to advocate consumption and production practices aimed to protect integrity of the environment. Furthering the paradigm, in 2005, Murphy pointed out that there was a need to not only positively treat the ecological environment but also demonstrate mindfulness toward the needs of the future generations. Advancing this discourse, Jamal and Jamrozy (2006) argued that sustainable marketing should aim to sustain the entire ecosystem instead of focusing on satisfying merely the needs and wants of people. In 2008, Donohoe and Needham proposed six fundamental tenets of sustainable marketing that integrated the three key pillars of sustainability (economic, environmental and social/cultural), emphasized on a holistic approach and the need to balance profit and consumer satisfaction goals with the overall well-being of the society. Enderle and Murphy called for including "a moral obligation in the sustainable development doctrine to promote a socially sustainable pattern of consumption" (Donohoe et al., 2016, p. 2).

Chhabra (2009) proposed a sustainable marketing model for heritage tourism in 2009 which featured eight inter-connected elements: mission, market segmentation, research, communication mix, local community involvement/benefits, partnerships and preservation/conservation. In 2010, she extended the model by adding six more sustainable components (local community involvement and benefits, partnership and collaboration, authenticity and conservation, visitor mindfulness, interpretation and economic viability) (Chhabra 2010).

Over the last decade, several scholars have tested and/or offered their perspectives on Chhabra's sustainable heritage tourism marketing model. For instance, Donohoe (2012) reported that the model offered value in the planning phases of the marketing process. While his study confirmed the significance of the partnership, preservation and conservation elements in shaping the core aspects and long-term viability of the marketing plan, he expressed concerns about the implementation potential of the model. In this regard, he shared Murphy's question about its operational potential in that "sustainable marketing requires that we spend an equal amount of time, if not greater amount of time and energy on what happens after the individual or business consumes the product" (Murphy, 2005, p. 190). According to Donohoe, "this requires additional upfront costs to better understand the market and its impacts on the natural, cultural and economic environments (environment analysis, market segmentation and research), so that marketing can be informed by operational realities and savings can be gained over the long-term" (2012, p. 139). Commenting on the SHTM model, Timothy (2021) suggested including a "set of practices that better support the principles of sustainability and the ideals of research and conservation. Practices such as partnerships, involvement of local communities, environmental analysis and conservation must underscore the marketing efforts of heritage sites as they continue in their efforts to educate, entertain and protect" (2021, p. 310).

Transformative travel is about travel experiences that capture 'journeys of the mind' aiming at self-fulfillment and responsible behavior toward humanity and natural and built ecosystems (Reisinger, 2013). As I mentioned in a previous paragraph, the pandemic has evoked fresh insights and ways of understanding and defining transformation so that it is not limited to peak/momentary glimpses of happiness (Chhabra, 2021b). It has stimulated inspection of ways in which transformations can last and be meaningful, mindful and sustainable, both to the self and to the society at large. Following Sheldon's (2020) semantic paper on transformation, Chhabra suggested a transformative paradigm that stresses on transcendence to inculcate the doctrine of sustainable behavior, at the demand and supply sides of the tourism system. She posited that the ultimate goal should be to embark on a journey toward a state of awakening and her paradigm suggested an intrinsic journey to attain a state of 'Being' that transcends the self to an elevated/ultimate state of consciousness. She pointed out that this path can be "initiated/integrated through niche tourism platforms and be guided by the transformational learning and existentialist theories to support a co-created flow of peak moments or B-cognition (by both disoriented feelings and positive emotions), invigorated by the suppliers, hosts, facilitators and the tourists. If peaked experiences or B-cognition moments reoccur and are cemented by a deep learning element and introspection, they can inculcate a sustained desire to continue or progress on the inner transformation journey" (2021, p. 767). Several studies are emerging that promote the notion of transformative travel and sustained transformation to attain a higher state of consciousness. Niche forms of tourism (such as heritage tourism), if marketed in a sustainable manner, hold potential to shape the well-being of tourist selves and the host environments. In the contemporary times, there is a need to revisit sustainable marketing paradigms for heritage tourism from a transformational standpoint.

This book extends this line of inquiry. It focuses on sustainable marketing of transformative heritage tourism and takes a unique approach by integrating elements associated with the promotion of heritage justice, morally motivated existentialist authentic experiences and well-being of the visited and visiting environments (Mkono, 2020). The focus is drawn toward the heritage of communities such as local, descendent and indigenous communities. Existentialist authenticity relates to being true and authentic to oneself (Chhabra, 2021a; Steiner & Reisinger, 2006; Wang, 1999). Ethical (moral) consumption processes should examine the potential of ideal negotiation of existentialist authenticity to align with the sustainable transformation of one's optimal state of relaxation, one which is guided by responsible behavior (Chhabra, 2020; Mkono, 2020).

In several cases, including a heritage justice agenda provides advocacy and voice to marginalized communities. A heritage justice standpoint can assist in examining issues of inequalities, injustices or violations of heritage and cultural rights. Heritage justice implies taking permission or offering fair compensation to use indigenous culture as part of a heritage tourism product. Humanistic-based heritage justice agenda needs to be integrated in authentically claimed heritage tourism offerings to ensure restoration, protection and promotion of human dignity, autonomy and improvement of well-being and entrepreneurial opportunities for the indigenous communities (Camargo & Vázquez-Maguirre, 2021; Chhabra 2012).

Furthermore, marketing pursued by different heritage institutions needs to be critically examined from a sustainability and well-being standpoint. For instance, museums are appropriately situated to promote heritage justice and community healing in areas of health, social security and social justice. Some existing studies have reported an association between 'collective efforts to preserve cultural continuity' and improved health and well-being and lower suicide rates, for instance, in the First Nations communities in Canada (Simpson, 2009). Several examples of museum initiatives can be noted such as facilitating organization of cultural camps in a number of indigenous communities in Canada and Australia to immerse young people in cultural experiences and ceremonial practices and inculcate a sense of pride.

Marketing strategies, inclusive of a sustainability agenda, can inspire responsible consumption of tourism resources and help to inculcate moral values and mindfulness among the visitors, toward the host community and the visited cultural and natural landscapes. The influence of macro environment factors on tourism policies and marketing agenda cannot be denied (Chhabra 2015). However, today, some trends are fading away and new trends are merging (Ateljevic, 2020). Undeniably, the COVID-19 has changed everything about marketing and an innovative fluid "mindset (rolling iterations based on quick-paced and changing marketing conditions), premised on sympathy, empathy, respect and trust" is required of tourism institutions (Chhabra, 2021b). A tourism systems approach toward niche tourism marketing, underpinned on synergies between different stakeholders of tourism, can lead to innovated and value-added initiatives. The goal of sustainable marketing should be to identify/develop alternative strategies that discourage exploitation of local heritage and consider the overall well-being of the society. A sustainable marketing plan should integrate social and selective marketing concepts and aim to enhance/promote adaptive capacity of the whole tourism system.

In summary, in the contemporary times, sustainable marketing needs to be guided by a multifold purpose which promotes heritage/social justice; conservation of heritage; safeguarding of local culture and cultural landscapes; moral selving of heritage tourists; eudaimonia-lasting intrinsic transformation of the guests and hosts; economic viability for the host and indigenous communities; and co-created mindful spaces. The transformative tourism notion stresses on the need to promote messages to transform the general public and the tourist mindset toward awakened consciousness to benefit the visited and visiting communities and the planet earth in general (Pung & Chiappa, 2020). For instance, taking a tourist standpoint, recent scholarship clearly suggests that tourism research should shift its focus to outcomes associated with existentialist experiences and self-fulfillment which continue to nurture eudaimonic well-being, upon returning home (post-trip phase) (Chhabra, 2022; Sheldon, 2020; Wolf et al., 2017). Additionally, several tourism researchers are directing the need, for cultural and heritage institutions, to focus on the whole human side of a tourist (Cavender et al., 2020; Sheldon, 2020; Teo et al., 2021). For instance, the theory of conscientization can be embraced to stimulate critical consciousness (Chhabra, 1991; Sivananda, 1960). According to Meyer (2003), inner transformation is about values and practices of people found in a place. It is rooted in the cultural consciousness of daily life. Positive emotions and flourishing state of mind can be promoted through mindful interactions, cultural immersions artistic and intellectual pursuits.

Furthermore, to accomplish the aforementioned goals, integration of a socially responsible component (rooted in the social marketing concept) into the sustainable marketing protocol has become critical in the contemporary times. The traditional marketing mix is not perfectly suited to the field of heritage tourism and marketing literature in the recent decades is stressing on a social marketing perspective. The social marketing concept was introduced by Kotler and Zaltman (1971) who suggested that marketing can be used as a tool to resolve social and health-related issues. Kotler and Zaltman defined social marketing as "the design, implementation and control of programs calculated to influence the acceptability of social ideas and involving considerations of product planning, pricing, communication, distribution and marketing research" (1971, p. 5). Although it was rooted in the traditional marketing mix, it called for attention toward social and ethical obligations of enterprises, both non-profit and for-profit. It emphasized on long-term consumer welfare and reminded that businesses should include ecological and social factors in their product development and marketing strategies. According to MacFadyen et al., "marketing can encourage people to adopt behaviors that will enhance their own- and their fellow citizens' lives" (1999, p. 1). Social marketing holds tremendous potential to shape the behavior, not just of ordinary citizens and tourists, but also of policymakers and different stakeholders of tourism. Social marketers can target messages to transform the behavior of the media, for-profit/non-profit enterprises and policy and lawmakers (Bloom 1980; Laczniak & Murphy, 2012; MacFadyen et al., 1999). Besides emphasizing on consumer orientation, exchange and a long-term approach, MacFayden et al. suggest a 'beyond the consumer' element. This implies that the goal is not only to shape the behavior of consumers but also of other stakeholders or role players of tourism (Chhabra 2004). These can be vendors, corporations, agencies, organizations, societies and the public sector. Social marketing conventionally seeks six types of social change which integrate short-term and long-term time frames, micro, group and macro levels of societies (Levy & Zaltman, 1975). The aim is not just to influence the consumers and other important stakeholders individually but also the environment in which they function (Lawther & Lowry, 1995; Lawther et al., 1997; Murray & Douglas, 1988).

Structure of the Book

This book is divided into twelve chapters. The introduction chapter focuses on the notion of sustainable marketing. Key functions of sustainability and sustainable development goals are summarized. The chapter emphasizes on integrating a socially responsible perspective of marketing grounded in moral and secular ethics. Next, these perspectives are examined in the context of heritage tourism. A synopsis is offered on the manner in which sustainable marketing has progressed in the field of heritage tourism. Building on the work of previous studies, it is argued that the core dimensions of marketing (mission, environment analysis, communication mix and mark segmentation) should be strategically guided by a well-defined sustainable marketing and ethical consumption ethos rooted in elements/concepts/theories specifically relevant to heritage sustainability. These features refer to authenticity, conservation, partnerships with special emphasis on the locals, economic viability, mindfulness/interpretation techniques and corporate social responsibility with an emphasis on code of ethics. The chapter suggests a guided sustainable marketing model for heritage tourism. It is argued that the model will continue to evolve under the influence of political interventions, sustainable development goals and technological innovations such as the Metaverse. Chapter 2 discusses the nebulous nature of heritage tourism and outlines key heritage tourism trends. A contemporary view of authenticity is scrutinized because authenticity is regarded as the cornerstone of heritage tourism. Next, the chapter delves deeper into the heritage tourism transformation phenomena and discusses key theories of and models of transformation. A conceptual path toward transformative heritage tourism is proposed.

Chapter 3 suggests that staycations hold tremendous potential in stimulating innovative domestic changes and can be tapped by heritage tourism planners and marketers to devise strategies for designing and promoting innovative and authentic tourism products. The author illustrates that staycations offer a novel way, to local residents, of exploring and rediscovering a new perspective to explore and rediscover their hometown and local heritage orchestrations. Personalized authenticity can be stimulated through storytelling programs, for instance. According to Xie, such programs are "especially important for heritage-based staycation where storytelling, as an integral part of marketing tools for new places, has gained tremendous popularity, serving an incentive to transform heritage tourism."

Chapter 4 by Lee et al. (2024) focuses on transformative aspects of tea tourism. The authors illustrate that tea heritage-related tourism holds potential to enhance the health, well-being and immersive experiences of tourists in addition to enhancing the subsistence of "tea communities, farmers and villagers in tea producing regions." Lee et al. (2024) illustrate that it is important to develop and market this niche form of heritage tourism in a sustainable manner so that it can cultivate transformative qualities. The case study is located in Northern Thailand. Chapter 5, by He, investigates satisfaction levels of tourists and their desire for an authentic experience by analyzing their comments on local homestay websites. The author reports that tourists who are seeking authenticity share concerns about physical comfort. It is noted that they are more likely to hold negative perceptions and attitudes toward the attraction "even when the discomfort constitutes the so-called 'authenticity' and tourists with low ratings for their accommodation experience were more likely to have negative attitudes towards the physical environment." In other words, the chapter demonstrates that tangible comfort can intervene and influence tourists' desire for authentic experiences and satisfaction.

Chapter 6, by Pintel and Chhabra, examines a creative engagement technique employed by museums: folksonomy and social tagging. Social tagging is the allocation of keywords to resources that results in Folksonomy which is the accumulation of concepts expressed in a socially constructed classification system. Taggers are able to utilize their personal connection and interpretation in order to define various artifacts from museums. This study makes an effort to determine if folksonomy is a useful mechanism to engage heritage audiences according to the latest generations: Generation Z and Generation Alpha. The authors also gather insights on whether or not social tagging is an abrogation of museum responsibility. Overall, the results indicate that social tagging is an important tool to engage the young generations in heritage museums. The authors suggest a strategic participatory approach, for museums, grounded in citizen science and citizen humanities theories to promote public well-being and sustainability by engaging people in artwork that conveys sustainable messages. Chapter 7 (by Sawhney et al.) calls for a regulatory-based mechanism toward tribal tourism to ensure that it is based on sustainable development and empowerment of tribes and recognizes the uniqueness of their heritage and cultural norms. The authors examine the historical context of tribal communities in India, and trace their struggle as they encounter the challenges posed by trends associated with modernization and globalization. They scrutinize the existing tribal tourism framework and "its impact on the socio-economic conditions of tribal groups, highlighting the gaps and inadequacies in current policies and practices."

Chapter 8, by Timothy, focuses on the use of heritage as a branding tool to develop a sense of place. The chapter examines the significant role of various forms of cultural heritage in building identification and branding, both spontaneously and in a planned manner. Brands at the international and regional levels are discussed based on the way they facilitate growth in tourism and foster solidarity in their respective regions. Prestigious designations such as the WHS peak the interests of countries to embrace tourism. Timothy illustrates that the national brands are also tapped in marketing strategies to stimulate growth and extend visibility toward the tourists and the tour agencies.

Chapter 9, by Yılmazdoğan and Ece, employs a rebranding and retro-branding perspective to suggest incorporation of elements that hold potential to preserve and promote heritage in a sustainable manner. The author suggests that rebranding or retro-branding can be utilized to modify the marketing strategies of a destination and this technique can significantly alter its image and appeal. The chapter investigates and assesses the "strategic applications of rebranding and retro-branding in the context of heritage destinations." Based on a review of documented literature, the authors "dissect the nuances of how historical, cultural, or natural heritage sites can harness these marketing strategies to enhance their appeal, preserve their integrity, and boost tourist engagement." Various case studies are examined and successful practices and pitfalls of the rebranding and retro-branding processes are identified. In doing so, the authors offer important information and suggestions for the heritage tourism stakeholders. This chapter aims to offer a comprehensive discourse on sustainable heritage management and marketing. It offers valuable information from the standpoint of academics, practitioners and policymakers who are looking for effective sustainable branding strategies in a progressive competitive tourism environment. Extending this discourse, Viana-Lora, in Chapter 10, looks at the significance of sustainable marketing in heritage tourism from the supply side. A systematic literature review and content analysis is conducted. It is noted that the management and promotion of heritage tourism present some major challenges, but sustainable marketing is recognized as the appropriate tool to certify the sustained preservation of cultural heritage and to decrease externalities of tourism on the ecological environment and host communities. Viana-Lora points out that "there is a need to advocate for strategic planning that reconciles the often-competing interests of various stakeholders and ensures effective management and promotion of cultural heritage tourism."

Chapter 11, by Fagence, warns of dissonance in the manner transformation is promoted and interpreted. The author argues that any scenario where a historical story is being narrated from a tourism lens will be subjected to multiple interpretations based on the circumstances that shape the story's practicality. Fagence indicates that different stakeholders can hold conflicting views of heritage and orchestration of several transformation paths can trigger shaded memories which can obstruct the path toward achieving a state of self-fulfillment at the site and at home upon return. It is pointed out that once the person returns home, the sustained impact of experience will be impacted by reminiscence of discordant chronicles. In other words, the authors argue that the consumption of transformative experiences can be a complicated phenomenon as it is likely to be impacted simultaneously by multiple connotations. In other words, "a comfort level between the story tellers, the setting and the tourist is required. Equally important is the legitimization of the story tellers and their stories and the level of trust the tourists place in them." The author's study is anchored in several backdrop areas of concern such as particular aspects of tourist's experience, unforgettable tourism encounters and the transformative theory.

In Chapter 12, I revisit the key messages communicated by the contributors in the context of sustainable marketing and transformation. Insights and emerging themes gathered from the work of recent scholars are shared, followed by a brief overview of some general tourism trends. Emerging perspectives of balanced authenticity and authenticity as a conduit for self-improvement and well-being are discussed. Notions of cultural self-improvement and heritage responsibility are touched upon. Next, challenges associated with Metaverse are discussed, followed by a reflection on the work of scholars focusing on the well-being and transformation of the host community. The chapter closes with the argument that sustainable marketing needs to be grounded in socially responsible codes guided by the Eastern ethical system of 'dharma' and 'karma' proposed by Sheth and Parvatiyar (2021). This path guides marketing to be planned and operationalized from an elevated state of consciousness and mindfulness, that is, from a eudaimonic state of mind. A transformative sustainable marketing doctrine is proposed. All chapters contribute to the overarching focus of the book in a robust manner. For instance, they focus on different aspects of sustainable marketing and heritage tourism, highlight emerging issues and discuss possible resolutions. This book endeavors to foster critical awareness of the need to design a socially integrated and guided sustainable marketing model that can open meaningful pathways toward transformative heritage tourism by aspiring toward holistic well-being of the self and the society through an elevated sense of consciousness (eudaimonia). All those who are interested in sustainable marketing, transformative heritage tourism and social, ecological, cultural and economic welfare of visited and visiting destinations will find this monograph educational and useful.

References

Ateljevic, I. (2020). Transforming the (tourism) world for good and (re) generating the potential 'new normal'. *Tourism Geographies*, *22*(3), 467–475.

Bloom, P. N. (1980). *Evaluating social marketing programs: Problems and prospects*. The 1980 Educators Conference Proceedings. American Marketing Association.

Camargo, B. A., & Vázquez-Maguirre, M. (2021). Humanism, dignity and indigenous justice: The Mayan train megaproject, Mexico. In *Justice and tourism* (pp. 239–258). Routledge.

Cavender, R., Swanson, J. R., & Wright, K. (2020). Transformative travel: Transformative learning through education abroad in a niche tourism destination. *Journal of Hospitality, Leisure, Sport & Tourism Education*, *27*, 100245.

Chhabra, D. (2004). Research note redefining a festival visitor: A case study of Vendors attending Scottish Highland Games in the united states. *Event Management*, *9*(1–2), 91–94.

Chhabra, D. (2009). Proposing a sustainable marketing framework for heritage tourism. *Journal of Sustainable Tourism*, *17*(3), 303–320.

Chhabra, D. (2010). *Sustainable marketing of cultural and heritage tourism*. Routledge.

Chhabra, D. (2012). A present-centered dissonant heritage management model. *Annals of Tourism Research*, *39*(3), 1701–1705.

Chhabra, D. (2015). *Strategic marketing in hospitality and tourism: Building a 'SMART' online agenda*. Nova Science Publishers, Inc.

Chhabra, D. (2020). Transformational wellness tourism system model in the pandemic era. *International Journal of Health Management and Tourism*, *5*(2), 76–101.

Chhabra, D. (2021a). *Resilience, authenticity and digital heritage tourism*. Routledge.

Chhabra, D. (2021b). Transformative perspectives of tourism: Dialogical perceptiveness. *Journal of Travel & Tourism Marketing*, 38(8), 759–768.

Chhabra, D. (2022). Transformational model of well-being for serious travellers. *International Journal of Spa and Wellness*, *5*(2), 129–146.

Chhabra, G. S. (1991). Journey of the Mind. Forthcoming Manuscript. University of Jammu, India.

Donohoe, H. M. (2012). Sustainable heritage tourism marketing and Canada's Rideau Canal world heritage site. *Journal of Sustainable Tourism*, *20*(1), 121–142.

Donohoe, H. M., Salo, B., & Gilmore, T. C. (2016). The Promise and Potential of Sustainable Heritage Tourism Marketing. TTRA International Conference. Travel and Tourism Research Association. https://scholarworks.umass.edu/

Jamal, T., & Jamrozy, U. (2006). Collaborative networks and partnerships for integrated destination management. In *Tourism management dynamics* (pp. 164–172). Routledge

Jolliffe, L., Piboonrungroj, P., & Chaikor, J. (2024). Transformative tea tourism in Northern Thailand. In D. Chhabra (Ed.), *Sustainable marketing of transformative heritage tourism*. Routledge. In Press.

Kotler P., & Zaltman G (1971). Social marketing: An approach to planned social change. *Journal of Marketing*, *35*, 3–12.

Laczniak, G. R., & Murphy, P. E. (2012). Stakeholder theory and marketing: Moving from a firm-centric to a societal perspective. *Journal of Public Policy & Marketing*, *31*(2), 284–292.

Lawther, S., Hastings, G. B., & Lowry, R. (1997). De-marketing: Putting Kotler and levy's ideas into practice. *Journal of Marketing Management*, *13*(4), 315–325.

Lawther, S., & Lowry, R. (1995). Social marketing and behavior change among professionals. *Social Marketing Quarterly*, *II*(1), 10–11.

Lee J., Piboonrungroj, P., & Chaikor, J. (2024). Transformative tea tourism in Northern Thailand. In *Sustainable marketing of transformative heritage tourism*, D. Chhabra (Ed.). Routledge. In Press.

Levy S. & Zaltman G. (1975). *Marketing, society and conflict*. Prentice Hall.

MacFadyen, L., Stead, M., & Hastings, G. (1999). A Synopsis of Social Marketing. Stirling: Institute for Social Marketing, 295. https://www.stir.ac.uk/about/faculties/health-sciences-sport/research/research-groups/institute-for-social-marketing.

Meyer, M. A. (2003). Hawaiian hermeneutics and the triangulation of meaning: Gross, subtle, causal. *Social Justice*, *30*(4), 54–63.

Mkono, M. (2020). Eco-hypocrisy and inauthenticity: Criticisms and confessions of the eco-conscious tourist/traveller. *Annals of Tourism Research*, *84*, 102967.

Murphy, P. E. (2005). Sustainable marketing. *Business & Professional Ethics Journal*, *24*(1/2), 171–198.

Murray, G. G., & Douglas, R. R. (1988). Social marketing in the alcohol policy arena. *British Journal of Addiction*, *83*, 505–511.

Pung, J., & Chiappa, G. (2020). An exploratory and qualitative study on the meaning of transformative tourism and its facilitators and inhibitors. *European Journal of Tourism Research*, *24*, 2–21.

Reisinger, Y. (Ed.). (2013). *Transformational tourism: Tourist perspectives*. CABI.

Sheldon, P. J. (2020). Designing tourism experiences for inner transformation. *Annals of Tourism Research*, *83*, 102935.

Sheth, J. N., & Parvatiyar, A. (2021). Sustainable marketing: Market-driving, not market-driven. *Journal of Macromarketing*, *41*(1), 150–165.

Simpson, M. (2009). Museums and restorative justice: Heritage, repatriation and cultural education. *Museum International*, *61*(1–2), 121–129.
Sivananda, S. (1960). *Vedanta for beginners*. The Divine Life Society.
Steiner, C. J., & Reisinger, Y. (2006). Understanding existential authenticity. *Annals of Tourism Research*, *33*(2), 299–318.
Teo, C. L., Tan, S. C., & Chan, C. (2021). Pedagogical transformation and teacher learning for knowledge building: Turning COVID-19 challenges into opportunities. *Canadian Journal of Learning and Technology*, *47*(4), 2–26.
Timothy, D. J. (2021). *Cultural heritage and tourism: An introduction* (2nd ed.). Channel View Publications.
Wang, N. 1999. Rethinking authenticity in tourism experience. *Annals of Tourism Research*, *26*(2), 349–370.
Wolf, I. D., Ainsworth, G. B., & Crowley, J. (2017). Transformative travel as a sustainable market niche for protected areas: A new development, marketing and conservation model. *Journal of Sustainable Tourism*, *25*(11), 1650–1673.

1

THE CORE OF SUSTAINABLE MARKETING

Deepak Chhabra

Over the past few decades, sustainability has surfaced as a crucial marketing perspective. This notion has been broadened to embrace socio-ecological and environmental matters in financial and economic and management endeavors. Recent trends call for a change in marketing focus from responding to consumer needs to channeling market needs toward conscientious receptiveness on sustainable products and development of sustainable societies (Laczniak & Shultz, 2022). In other words, it has become crucial to amend the marketing orientation from merely being responsive to consumer needs into developing a more responsible approach to drive consumer markets to embrace sustainable products and services and build responsible societies (Villena & Gioia, 2020; Winston, 2019).

First global awareness of the pitfalls of unsustainable growth was presented in the report "The Limits of Growth for the Club of Rome's Project on the Predicament of Mankind" (Meadows et al., 1972). It stated that the exponential growth of resources was needed to avoid crash of the global system due to unchecked economic growth. The outcome of this awakening call led to research initiatives by several scholars to highlight/contribute toward socio-ecological prospects of marketing (Cravens, 1974; Fisk, 1973; Henion, 1976; Perry, 1976; Shapiro, 1978; Sheth & Wright, 1974). The Brundtland Report by World Commission on Economic and Development (WCED) initiated the need to integrate "economic development, natural resource management, and social inclusion and equity." The Commission defined sustainable development as development "to ensure that it meets the needs of the present without compromising the ability of the future generations to meet their own needs" (WCED, 1987, p. 8). The report guided the key discussions at the Earth Summit in 1992. According to the United Nations (UN), development

DOI: 10.4324/9781003346197-1

TABLE 1.1 Key Functions of Sustainability

1	Key Functions of SustainabilityMake optimal use of environmental resources that constitute a key element in tourism development, maintaining essential ecological processes and helping to conserve natural heritage and biodiversity.
2	Respect the socio-cultural authenticity of host communities, conserve their built and living cultural heritage and traditional values, and contribute to inter-cultural understanding and tolerance.
3	Ensure viable, long-term economic operations, providing socio-economic benefits to all stakeholders that are fairly distributed, including stable employment and income-earning opportunities and social services to host communities, and contributing to poverty alleviation.
4	Promote strategic partnerships and mutually beneficial collaboration between the stakeholders.
5	Ensure strong leadership is in place to confirm organized progress to achieve the goals.
6	Promote mindfulness in tourists and ensure they have a meaningful experience which has a potential to transform them.

Source: UN-WTO (2017).

of sustainable tourism calls for a harmonious balance between environmental, economic and socio-cultural pillars of sustainability (see Table 1.1). Sustainable development goals (SDGs) were founded on three areas of consideration: economic, social and environmental pillars of sustainability.

Sustainable Development Goals

The United Nations SDGs are deemed to be the most feasible route to guide sustainable development and promotion of tourism. Approximately 159 countries have endorsed the SDGs. Through the SDGs, the UN has particularly demonstrated its commitment to advance knowledge and tourism protocol for promoting responsible tourism and using tourism as a conduit to promote viable economic growth, equitable development and ecological sustainability (United Nations, 2023). The UN continues to play a key role in facilitating implementation of "global code of ethics" in the tourism industry to optimize socio-economic benefits of tourism and reduce negative impacts.

It is important to note that the SDGs cannot be accomplished without the pivotal interventions at the corporate and government levels. Despite rising ethical consumption consciousness of consumers, Sheth and Parvatiyar (2021) argue that a market drive by itself is not adequate to safeguard sustainability at the grassroots level. This implies that, "more conscious and conscientious institutional interventions are needed" (Sheth & Parvatiyar, 2021, p. 151). The SDGs, in this regard, can shepherd in developing a strategic path. The 2030 Agenda for SDGs was developed by the UN to ensure initiatives are taken to safeguard the planet and all living beings and many countries have become signatories to it. Tahiri et al. note that "as the

17 SDGs offer the world a new direction, such as living and working in a sustainable way, they can be applied in various ways in our everyday activities" (2021, p. 1). From a generic standpoint, SDGs have set a broad spectrum of universal targets for viable environment, society and economy. They also offer opportunities to develop cutting-edge products and services that adhere to some or most of the SDGs and help align commercial goals with sustainability principles to enhance brand awareness and proffer certified privileges for operation (McClements et al., 2021).

A core tenet of sustainability is to ensure promotion of an ethical lifestyle that is mindful of the natural environment and other tangible and intangible resources. Tahiri et al. (2021) write that such initiatives are "beneficial for human health and well-being in addition to socio-economic benefits and contribute to environmental integrity. We all have to be obligated to save the planet in which we live throughout our daily and local activities" (p. 1). The SDGs offer numerous paths to help achieve a better future and community without jeopardizing the key pillars of sustainability. The 2017 International Year of Sustainable Tourism for Development highlighted the important role tourism can play in advancing SDG12 (Responsible Consumption and Production), SDG8 (Decent Work and Economic Growth) and SDG14 (Life Below Water) (UNWTO, 2018). Attention is now needed to mobilize the SDGs in an effective manner to ensure sustainable guidelines are adhered (Boluk et al., 2017, 2021; Bramwell et al., 2017; Hall, 2019). Today, the SDGs have become the focal point of reference to span the contribution of tourism toward sustainable development (Ali et al., 2020; Boluk & Rasoolimanesh, 2022; Boluk et al., 2021; Hall, 2019; Rasoolimanesh et al., 2020; Scheyvens & Cheer, 2021; UNWTO and UNDP, 2017). A comprehensive list of the seventeen SDGs is presented in Table 1.2.

With regard to the enterprise role in sustainable development, the SDGs clearly convey a message to all businesses to harness their innovative and creative strategies to solve issues that inhibit sustainable development (Varadarajan, 2019). In other words, the enterprises are expected to devise business-driven solutions and design/use technologies to accomplish these goals/targets to advance all societies across the globe. The 2023 Sustainable Development Goals Report (United Nations, 2023) presents a forthright appraisal of the SDGs based on recent data. The report outlines gaps and urges the world to increase its initiatives by garnering strong political support and tapping into technological innovations, resources and information. Collectively, the community can inspire progress toward accomplishment of the SDGs and strive toward a promising future for everyone. The report states that the globe has arrived into the age of "polycrisis" and highlights the devastating impacts of the war in Ukraine, poor world economy and prolonged effects of COVID-19 in impeding the progress toward the SDGs. In fact, the globe is still recovering from the effects of COVID-19 such as reverse progress in global health and

TABLE 1.2 United Nations List of Sustainable Development Goals (SDGs)

	Theme	*Description*
SDG1	Poverty	End poverty in all its form
SDG2	Zero hunger	End hunger, achieve food security and improved nutrition and promote sustainable agriculture
SDG3	Good health and well-being	Ensure healthy lives and promote well-being for all at all ages
SDG4	Quality education	Ensure inclusive and equitable quality education and promote lifelong learning opportunities for all
SDG5	Gender Equality	Achieve gender equality and empower all women and girls
SDG6	Clean water and sanitation	Ensure availability and sustainable management of water and sanitation for all
SDG7	Affordable and clean energy	Ensure access to affordable, reliable, sustainable and modern energy for all
SDG8	Decent work and economic growth	Promote sustained, inclusive and sustainable economic growth, full and productive employment and decent work for all
SDG9	Industry innovation and infrastructure	Build resilient infrastructure, promote inclusive and sustainable industrialization and foster innovation
SDG10	Reduce inequalities	Reduce inequality within and among countries
SDG11	Sustainable cities and communities	Make cities and human settlements inclusive, safe, resilient and sustainable
SDG12	Responsible consumption and Production	Ensure sustainable consumption and production patterns
SDG13	Climate action	Take urgent action to combat climate change and its impacts
SDG14	Life below water	Conserve and sustainably use the oceans, seas and marine resources for sustainable development
SDG15	Life on land	Protect. Restore and promote sustainable use of terrestrial ecosystems, sustainably manage forests, combat desertification, and halt and reverse land degradation and halt biodiversity loss.
SDG16	Peace, justice and strong institutions	Promote peaceful and inclusive societies for sustainable development, provide access to justice for all and build effective, accountable and inclusive institutions at all levels
SDG17	Partnerships for the goals	Strengthen the means of implementation and revitalize the global partnership for sustainable development

Source: United Nations (2023).

huge economic losses. According to the report, "the climate crisis is worsening as greenhouse gas emissions continue to rise. Catastrophic and intensifying heat waves, droughts, flooding and wildfires have become far too frequent. Rising sea levels are threatening hundreds of millions of people in coastal communities. In addition, the world is currently facing the largest species extinction event since the dinosaur age and the oceans are burdened with plastic pollution" (United Nations, 2023, p. 2). There is a call for a structural shift in dedication, camaraderie, investment and action and correcting the historic injustices – to move the world toward a restored path.

Sustainable Marketing

Sustainable marketing, as a corporate practice and as a scholarly line of inquiry, has become a cornerstone in the contemporary era (Hult et al., 2018; Hunt, 2011; Kramer, 2020; Varey, 2010; Winston, 2019). Numerous factors can be attributed to this emerging trend. It is evident that "firstly, ... scientific evidence ... with people's experiential observations that climate change, environmental degradation, and effects of substantive economic inequalities around the world are not anymore a distant future; Secondly, many state governments ... at local, sub-national, and national levels, have made sustainability commitments to affect behavioral changes and economic activities that impact the environment and society" (Sheth & Parvatiyar, 2021, p. 150). Several initiatives can also be noted at the corporate level. For instance, Microsoft, Walmart and Procter & Gamble have taken initiatives and developed programs to engage important stakeholders to build supply chains to embrace social and environmental code of ethics (Bhattacharya & Polman, 2017; Stevens, 2019; Villena & Gioia, 2020; Winston, 2019).

Sheth and Parvatiyar (2021) present a sustainable marketing model that takes a shift away from the market and is propelled by the sustainability ethos. They call for a guided marketing approach toward regulatory mediation. Sheth and Parvatiyar note that although the "governments actively engage in translating policy commitments and sustainability priorities, including legally binding conventions, corporate marketers have to embrace transformative strategies to change business practices, particularly those linked to the unsustainable production, distribution, and consumption systems" (2021, p. 151). Policy interventions are required to ensure that sustainability goals are integrated into overall mission/vision of enterprises. An active persuasion toward sustainable behavior is imperative. The interventions at the business and government levels need to be synchronized for successful implementation of SDGs. While it cannot be denied that the contemporary trends indicate shift toward conscious consumption by consumers, the consumers are becoming more mindful toward the ecological environment systems and host communities they visit (Balderjahn et al., 2018; Kolar & Koklič, 2018; Luan & Chen, 2021). However, Sheth and Parvatiyar (2021) argue that these initiatives do not suffice to spark a major shift toward sustainability and make a plea for more mindful and devoted interventions at the institutional level (Hult et al., 2018; Laczniak & Shultz, 2021; Winston, 2019).

It is also being noted that investment companies, on the world stage, now require corporations to provide evidence of compliance toward sustainability goals. This is likely to promote more adherence to responsible practices and ethical considerations (Kramer, 2020). Macro perspectives of sustainability are also fetching attention (Laczniak & Shultz, 2022; Sheth et al., 2011; Varadarajan, 2017). Unavoidably, the macro environment factors play a pivotal role in shaping consumer behavior and enterprise strategies toward sustainable marketing. Political, economic, social and technological (PEST) analysis has been frequently used to evaluate the macro-environment effects on organizations (Chhabra, 2015; Kotler et al., 2016; Timothy, 2021; Tleuberdinova et al., 2021). Political factors commonly relate to legislation and policies, taxation systems and political stability. Economic factors include the current state of economy, interest rates, currency exchange rates, prices/inflation, taxation laws, financial regulatory environment, ability to secure loans, distribution of wealth in a society, globalization and type of business ownership such as vertical and horizontal integrations. Social factors include demographic shifts and social concerns and type of lifestyles such as affluent, healthy and marginalized. Technological factors usually allude to product, communication and operational technologies. Besides PEST, other important factors that can impact planning, development and management strategies of tourism businesses include climate/natural, geographic location/topography and cultural. These can differ based on different geographic spaces. Attributes at the destination level can prompt, for instance, entrepreneurship initiatives. The destination life cycle stage can also play an influential role in molding strategies (Weiermair et al., 2007).

Several studies have made a dedicated effort to illustrate the manner in which marketing can play a key role in promoting sustainable communities and corporate practices (Chhabra, 2010, 2021; Mittelstaedt et al., 2014; Polonsky & Rosenberger, 2001; Prothero & McDonagh, 2015). However, proactively integrated political interventions, in corporate marketing practices, are critical to reinforce progressive commitment toward sustainability (Sheth & Parvatiyar, 2021). The notion of aggregate marketing system (AGMS) stresses on the shared role of marketers, consumers and public agencies in promoting the societal aspects of marketing (Sheth & Parvatiyar, 2021; Wilkie & Moore, 1999). The responsibility rests on the shoulders of corporate marketers to design transformative strategies and stay assiduously involved in metamorphizing the market mindset toward sustainable behavior.

Role of Enterprises in Promoting Sustainable Development

Emphasis on the role of enterprises in facilitating and building a more socially and environmentally mindful business ecosystem has become more pronounced today (Tahiri et al. 2021; Villena & Gioia, 2020). Recent studies call for enterprise sustainability to move beyond financial and competitive sustainable endeavors to target societal and ecological sustainability (Vila & Bharadwaj, 2019). According to Varadarajan (2019), corporate sustainable initiatives need to focus on triple bottom line of people, planet and profits. This implies that capitalism needs to be transformed by changing systems thinking to cordon inequitable growth. This calls for expanding the breadth and pace of sustainable marketing initiatives (Sheth & Parvatiyar, 2022) by dispelling stereotype perceptions of having to choose between sustainability and profits by moving beyond the conventional business mindset.

Some organizations are developing a competitive advantage and exerting their buying authority by requiring supply chain stakeholders to adhere to sustainability goals and practices (Bhattacharya & Polman, 2017; Villena & Gioia, 2020; Whelan & Fink, 2016). At the enterprise level, focus is on the "entire value chain through which sustainability goals can be impacted" (Laczniak & Shultz, 2022, p. 155). As mentioned earlier, recent literature calls for the corporate sector to embrace a more obligatory role in facilitating sustainable planning and behavior (Mitchell et al., 2010; Shultz & Holbrook, 1999; van Dam & Apeldoorn, 1996; Varey, 2012). Congruently, it is important to recognize the significant role of government in steering the consumer demand, toward ethical and sustainable consumption, by influencing their needs/wants. The marketing professionals can take initiatives to understand the consumer black box (Lin & Ding, 2005). Some scholars argue that the consumer black box can be influenced to initiate more responsible and sustainable behavior. It is in the consumer black box that, "marketing stimuli such as the marketing mix and environment factors (for instance economic, social, political, etc.) are converted into an array of noticeable travel-related responses" (Chhabra, 2012, p. 344) such as selection of a travel destination, timing of purchase and spending amount-related decisions (Kotler et al., 2016). The box comprises two core elements: attributes of travelers (e.g., social, personal, cultural and psychological) and their decision-making process beginning with need recognition, actual purchase and cognitive dissonance (Kotler et al., 2016).

Alternative marketing approaches are also necessary to attribute similar values to "nature, planet, and ecological sustainability as sources of well-being for humans" and other living beings (Borland & Lindgreen, 2013). According to Seyfi et al., "Covid-19, in a way triggered transformation of the tourism industry and has nudged it towards more receptivity and sense of responsibility and sustainable tourism" (2020, p. 343). Mobility and consumption patterns have been immensely impacted (Seyfi et al., 2020; Tandrayen-Ragoobur et al., 2022). The pandemic brought globalization to a standstill. Consequently, a shift toward accessible and proximate destinations has become a new trend and is the outcome of restrictions imposed on cross-border mobility. Xie (2024) reiterates this point in Chapter 3. The next section deliberates upon the notion of social marketing and talks about the significance of integrating a socially responsible doctrine in marketing.

The Notion of Socially Responsible Marketing

Several studies have inspected propaganda of enterprise ethics and socially and environmentally responsible strategies from a socially responsible marketing standpoint (Laczniak & Shultz, 2022). As pointed out by Ferrell and Ferrell (2022), "good corporate citizenship, stakeholder orientation and social/ecological sustainability" highlight key factors that can be employed to develop a corporate culture and operational tools to implement ethical marketing applications. Evidently, a harmonious blend at the micro and macro stages is required. Micro level alludes to the enterprise's internal environment, whereas macro refers to the community system as a whole, shaped by macro environment factors. According to Laczniak and Murphy (2018), both normative ideology and social science strands play a key role in knowing and strengthening marketing ethics.

Furthermore, a moral lens is essential to promote economic equity and distributive justice. Santos points out that justice-related issues have to shift from "marketing for" toward "marketing with" to ensure that the impoverished populations are treated in a fair manner (Santos & Laczniak, 2009). In the context of heritage tourism, sustainability initiatives need to circumscribe ecological, cultural, social and economic perspectives. According to Laczniak and Santos, the UN Goals for Sustainable Business Development "go far beyond ecological concerns in their aspiration, and include working conditions, economic growth, the nature of responsible production and poverty levels" (2022, p. 555). There is always a risk of "ethical walkback" when these practices compromise profit-driven goals (Laczniak & Santos, 2022; Nill et al., 2020).

Recent studies call for integrating socially responsible marketing deeply into the sustainable marketing philosophy. Sheth and Parvatiyar (2022) have examined the concept of socially responsible marketing from a "dharma" ethic standpoint. The authors embrace the Aristotelian notion of eudaimonia striving toward a wellness-centered and thriving lifestyle while simultaneously establishing connections at the macromarketing level. Drawing on this perspective, the eastern notions of "dharma" and "karma," this chapter argues that the same can foreground delivery of sustainable marketing and ultimately aim at the well-being of all stakeholders and the planet at large. Describing the work of Laczniak and Santos (2011), Sheth and Parvatyar write that from a societal obligation vantage point, "since 'well-being' and 'human flourishing' are presumably what all persons desire; marketing, being the great economic provisionary, should directly contribute to those outcomes. Aristotle's concept of Eudaimonia, which encompasses quality-of-life and societal well-being, can be envisioned as the ultimate purpose of marketing, ethically rendered" (2022, p. 556). Certain virtues need to be championed and practiced to achieve a eudaimonic state of mind. It is proclaimed that "universal virtues of righteousness will lead to cosmic and community harmony and balance" and since virtues can be manifold, their application relies on the nature of scenarios. The possible virtues are many, but the application of specific virtues needs to be situation dependent. According to Sheth and Parvatiyar, enduring marketing ethics should be anchored in the fundamental principles of "corporate citizenship, stakeholder orientation and social/ecological sustainability" (2022, p. 19).

Sheth and Parvatiyar (2022) further write that perfection in marketing should be examined as a normative macro-level notion that aims for affective/introspective outcomes associated with well-being. Sustainable marketing should be grounded in moral values and secular ethics penciled from philosophies such as occidental and oriental. The "explication of abiding values" is important because in the absence of enduring attributes such as care, empathy, altruism and comradeship, SRM will not be worthy and lack a moral barometer (Sheth & Parvatiyar, 2022). Along similar lines, Shapiro (2022) postulates that a discourse on socially responsible consumption (SRC) is important to refine the SRM perspective. Laczniak and Shultz write that "SRC is the other half of the marketer-buyer exchange and looking at each element of SRM primarily from the consumer and/or consumption perspective would be a valuable exercise. Such a dyadic approach may uncover incremental insights into the dynamics of socially healthy macromarketing systems" (2022, p. 558).

In essence, there is a dire need to develop a comprehensive outlook, given the complex nature of human life. Strategic initiatives should aspire to promote "informed, cooperative, coordinated, sustainable and thus socially responsible markets, marketing, public policy, and consumer behavior; this systemic activity moreover requires proactive, constructive engagement, which must occur locally, globally, and over time and space" (Laczniak & Shultz, 2022, p. 558). While recognizing the crucial significance of SRM, the next section deliberates upon the notion of sustainable marketing particularly in the context of heritage tourism. Furthermore, it argues that the field of sustainable marketing is like a big umbrella and its successful implementation, in the contemporary era, required integration of the SRM doctrine and collaborative ethos. The ethical codes associated with SRM can chaperone a meaningful path that inspires sustained transformation by cultivating responsible production and consumption practices.

Sustainable Marketing in the Context of Heritage Tourism

Initially, documented literature on heritage tourism marketing honed on commercial aspects of marketing such as market segmentation, consumer behavior, communication and promotion, and partnerships. A plethora of studies concentrated on identifying profitable target markets for heritage tourism so that programs and experiences could be tailored toward them. Market segments were identified based on demographics, motivations, activities, benefits and spending propensity (Chhabra, 2009; Kotler et al., 2016). Of late, experiential consumption has become an important delineating factor to bracket market segments based on type of experiences sought (Shobeiri et al., 2016; Ulusoy, 2016). Visitors seeking experiential consumption are also noted to have a high spending propensity. Evidently, in the earlier decades, the heritage tourism marketing initiatives were centered on developing promotional designs and message content aligned with selected or suggested target markets with profitability being the ultimate goal. The sustainable goals of heritage tourism marketing were mostly remiss. For instance, considerations toward authenticity, conservation, mindfulness, key stakeholder collaborations and viable economic benefits received scant attention. Having said that, some initiatives were noted in the early 21st century. For instance, the work of scholars like Misiura (2006), Rowan and Baram (2004) and McLean (2002) aided awareness toward sustainable aspects of marketing.

Although Misiura (2006) focused on consumers, her heritage marketing model acknowledged the influence of macro environment factors on heritage suppliers and the marketing mix. The key message conveyed by her model was the significance of fostering relationships to ensure/promote quality of service and consumer satisfaction. Misiura called on marketers to perform the role of ethnographers to obtain in-depth insights on consumers. She was of the view that heritage connects audience to consumer brands and community assets, thereby facilitating emotional and spiritual experiences. She emphasized on heritage-identity connections and heritage equity particularly in the context of showcasing the heritage of marginalized populations. She directed attention toward nostophobia and wrote that effort should be made to showcase all pasts.

In their seminal work on marketing archaeology, Rowan and Baram's (2004) work centered on archeology and consumption of foregone times. The authors held a stigmatized view of marketing and shared concern over the risks of commodifying archaeology and the past. In fact, they were skeptical of the term "heritage." They referred to power dynamics to decipher patterns in commodity production, distribution and consumption. Rowan and Baram opined that "the commodification of the past is part of a trajectory on which more and more aspects of social life and localized resources become objects for consumption" (2004, p. 6). The authors proclaimed that marketing taps on the past with a specific intent and that heritage tourism is a consumerist phenomenon. They pointed out that the addition of tourism (an economic aspect) in the dynamics distracted from the key purpose of archaeology, thereby underlining the gap between academic and the corporate agendas.

McLean's (2002) work centered on museums and she scrutinized marketing strategies and notions in heritage museum settings. The author argued that marketing should be regarded as a tool to accomplish museum goals associated with collection and preservation. According to her, "marketing at its lowest denominator is about building a relationship between the museum and the public" (McLean, 2002, p. 2). The author expressed concern over the manner in which the marketing term was misinterpreted by museums. McLean was also skeptical about the credibility of marketing with its multiple meanings and opined that it was skewed toward consumer desires and that supplier strategies centered on gratifying those needs. For the author "without people, there would be no need for marketing" (2002, p. 1). The author envisaged that the museum marketing ethos, based on public/social benefits, compromised its core reason of existence (preservation and authentic heritage/historic collections). Undoubtedly, museology is a complex notion and the ideologies of the 21st century expect museums to align with the societal values and integrate them in their interpretive content (Chhabra, 2008; McLean, 2002). Marketing is a conduit to ensure adaptation toward emerging trends which call for motivating messages to persuade the public to visit museums. McLean posited that "marketing needs should be specifically developed for the museum context, one which reflects the purpose of the museum" (2002, p. 2). She indicated that although marketing can trigger misapplications and misrepresentations, it can still perform a useful function if employed in a responsible manner. In the following years, several studies emerged which concentrated on the prominent aspects of sustainable marketing (such as partnerships/collaborations, conservation, authenticity and host community perceptions and benefits) albeit in an isolated manner. Only a handful of studies have appeared that take a somewhat holistic view of sustainable marketing in heritage tourism. The next section offers insights on some key sustainable marketing models in the context of heritage tourism.

Modeling Sustainable Marketing

From a generic standpoint, as mentioned earlier in this chapter, sustainable development is about satisfying the needs of the present generation without encroaching the aptitude of future generations to satisfy their needs. It offers a competitive advantage and its main goal should be to develop a blueprint for businesses to pursue sustainable prototypes of business. The triple bottomline approach to sustainability in the past few decades has stressed on enterprises to focus on accomplishing environmental and social justice in addition to striving for economic prosperity (Pomering et al., 2011). Pomering et al. (2011) developed a sustainability tourism marketing model by integrating (1) first the four Ps model (product, price, place and promotion) (McCarthy, 1960); (2) the services marketing mix model which includes three more Ps (participants, process and physical evidence) (Booms & Bitner, 1981); and (3) three additional Ps suggested by Morrison (2010) (partnership, packaging and programming). Pomering et al. (2011) postulated that all 10 Ps are related to each other within the context of the three sustainable development pillars and should be employed to build a sustainable tourism marketing model.

Few sustainable marketing frameworks can be identified that are devised exclusively for heritage tourism. Although Du Cros's (2001) model focuses on sustainability of heritage tourism, however, it offers useful implications from marketing standpoint. The author suggests a conservation-commodification scale and presents a matrix to demonstrate an evolving association between site robusticity and market appeal. On the matrix, "heritage places can be plotted based on their appeal to tourists and their robusticity or ability to withstand high levels of visitation" (Du Cros, 2001, p. 168). Positioning based on conservation/robusticity status offers significant insights to managers who are fixated on conserving cultural values for a longer period of time. The market appeal end of the scale provides noteworthy information from a tourism standpoint as it helps to evaluate worthiness of the heritage attraction based on market demand. Site-specific indicators are required to measure the positioning point on the continuum to guide protocol with regard to the level of permissible tourism development and conservation strategy/management of a heritage site. Du Cros also stresses on the importance of taking the political environment and host community views into consideration.

The Sustainable Heritage Tourism Marketing (SHTM) model proposed by Chhabra (2009), based on an empirical study of a purposeful sample of heritage museums across the United States, comprises mission statement, market segments, communication mix (interpersonal and impersonal channels), micro and macro environment factors (that are likely to influence the communication mix), partnerships and ongoing research initiatives. Chhabra (2009) shares that these influencing agents inform and guide museums in developing strategic marketing plans. Chhabra recommends that "marketing decisions need to be guided by conservation and preservation principles that underpin the traditional role of museums in addition to local community involvement and benefits" (2009, p. 313). A year later, Chhabra (2010) extended this model and tested it on different types of heritage attractions across the globe such as heritage sites, museums, heritage hotels, festivals and souvenirs. The amended SSHTM model (strategic sustainable heritage tourism marketing model) featured six more sustainable components (local community involvement and benefits, partnership and collaboration, authenticity and conservation, visitor mindfulness, interpretation and economic viability) in the outer layer. The inner layer comprised the foundational aspects of marketing: mission, communication mix, market segmentation, research and environment analysis.

Chhabra put forward a further extension to the SSHTM paradigm by emphasizing that engagement with other enterprises is critical to build successful partnerships while concurrently maintaining the exclusive and unique attributes of their heritage product at the grassroots level. That is, co-created enterprise engagement and exclusive management of unique attributes should both be regarded essential for sustainable marketing. With regard to the local community, civic engagement was deemed critical to facilitate mutually beneficial relationship between enterprises and the local residents. Strategies to promote ethical consumption of heritage and its support services were also championed for successful application of the sustainable marketing paradigm. Evidence also reiterated promotion of objective and existentialist versions of authenticity to ensure the tourist experiences were engaging and satisfactory. Several studies have lent credence to Chhabra's models (Donohue, 2012; Olya et al., 2018; Timothy, 2021).

Review of more recent literature on heritage tourism reveals a somewhat fragmented focus on sustainable marketing. For instance, scholars have focused on one aspect of sustainable marketing such as authenticity, interpretation, conservation and stakeholder collaborations with particular emphasis on host communities and co-created initiatives. Some new components associated with sustainable marketing can also be penciled such as cultural identity, emotions (Nowacki & Kruczek, 2021), nostalgia and preservation, prospects of embracing a CSR approach (Chhabra, 2015; Chi et al., 2019; Horng et al., 2018), social value of heritage and, technology/ICT particularly in the context of Metaverse, virtual reality (VR) and augmented reality (AR) (Buhalis et al., 2023a,b).

In summary, sustainable marketing needs to integrate a holistic systems approach and should be designed based on present and future needs and behavior/aspiration of consumers (Cuculeski et al., 2016). Simultaneously, as recent studies assert, they should also strive to shape human behavior of both the consumers and the stakeholders. As an instance, Cai et al. (2021)

postulate that collaborative development of cultural heritage protection and tourism should be regarded as a crucial aspect of sustainable heritage tourism. Using Mount Lushan (China) as a case in point, the authors illustrate that the mismanagement of the site was due to the failure in designing combined strategies associated with heritage protection and heritage tourism marketing. The authors suggest the use of digital technologies to overcome these shortcomings. According to Cai et al., "oblique aerial photography, 3D laser scanning technology, and 360 degrees panorama technology can digitize the Mount Lushan cultural landscape heritage site, integrating all elements to create a virtual tourism subsystem. It provides users with a virtual experience of cultural landscape heritage tourism and promotes cultural landscape tourism marketing. In addition, the tourist flow and the environmental subsystems were built through the integration of Internet of Things (IoT) technology and analytical models" (2021, p. 1). In this way, a tourist flow subsystem can be developed to assist management in organizing the flow of tourists guided by carrying capacity parameters. A pressure-state-response model is also recommended to assist the managers in assessing and managing environmental health.

In line with the foregoing, scientific tools are wanted to encourage stakeholders to persevere for SDGs so the tourism can be planned and administered in a sustainable manner. According to Passafaro (2020), there is a need to design all-embracing sustainable dispositions to aid tourism development, given that tourists are not a monolithic market segment but hold different travel preferences and personalities. For instance, backpackers are more mindful toward visited destinations than other types of travelers (Agyeiwaah et al., 2020).

Mousazadeh et al. (2023) use underground heritage tourism site (PQ-Persian Qanat) in Iran to illustrate that in-sustainable behaviors can generate negative impacts, particularly in fragile areas. The Qanats are a distinct manifestation of prehistoric Iranian technical innovation for use of groundwater and hold potential to develop into a novel tourist attraction. To ensure the preservation of the technical, cultural and historical values of PQs, the authors propose a new function to ensure the protection of PQs: developing sustainable behavior. They recommend development of the peripheral areas around the Qanats to ensure steady flow of tourism revenue which can then be used to protect the Qanats while opening them for tourism in a more regulated manner. Another highlighted element is sustainable mobility. Zones (car-free) for pedestrians and cyclists can help to build safe and environment-friendly spaces.

Sustainable mobility can offer a multitude of benefits for the PQs. The tunnels provide a safe and comfortable mode of transportation, shielded from the harsh weather conditions and extreme heat of the desert environment. The electric vehicles can decrease air pollution and carbon emissions and this can be accomplished by effective interpretation techniques and by raising awareness of the local community "about the importance of the Qanats and the significance of protecting and preserving them .... Sustainable practices such as sustainable footprint, reduction in tourists' noise pollution, sustainable mobility, buffer zone controls, and sustainable routing can ensure the sustainable protection of PQ heritage by shaping the behavior of tourists, officials, and natives" (Mousazadeh et al., 2023, p. 13).

From the foregoing, it is clear that a plethora of studies underline one or more aspects of sustainable tourism marketing and these elements offer important insights on how to alter existing models and make them relevant to the contemporary times. Worthy of note is that extant literature unanimously acknowledges the crucial role of digitalization in the design and implementation of sustainable marketing strategies. Next section discusses the increasing significance of integrating digitalization in marketing strategies.

Digitalization

The Internet has deeply proliferated the travel and tourism industry. In fact, the industry has been revolutionized by massive networks and smartphones which have built a smart infostructure (Buhalis et al., 2020). According to several scholars, technology has both disrupted and benefited the travel and tourism industry. For instance, it has directly connected the consumers to destinations, the lodging industry, airlines, etc., eliminating the middlemen. Shifting consumer attention toward firm-owned mobile app activities, sponsored/owned by the firm, can augment spending propensity and repeat behavior (Jang et al., 2021). Furthermore, the arrival of Global Distribution Systems (GDSs) (such as Amadeus and Sabre) has provided user-friendly interfaces to travel agencies to assist in obtaining information from a broad spectrum of suppliers and make consumer reservations for flights, hotels, railways, etc. (Aamir & Atsan, 2020). Contextual relevancy is galvanized by smartphones promoting meaningful value co-creation in real-time (Buhalis et al., 2020). The advent of 5G+ mobile networks and the Internet of Everything have enabled multi-dimensional (such as 3D) simulations and empowered the performance of immersive technologies (e.g., mixed reality, AR and VR) (Loureiro et al., 2020; Rauschnabel et al., 2022). According to Buhalis et al., these offer "unprecedented opportunities for tourism management and marketing as well as for engaging with multiple stakeholders immersively. Tourists can preview tourism destinations and services before purchasing whilst they can complement real experiences at the destination" (2023a,b, p. 2).

At the same time, advanced technologies pose threats because of the emergence of sharing economy and surging popularity of Airbnb, Uber, etc. (Chang & Sokol, 2022; Guttentag & Smith, 2017). Shifting consumer attention toward firm-owned mobile app activities, sponsored/owned by the firm, augments spending propensity and repeat behavior (Jang et al., 2021). Conservation of cultural and heritage resources is crucial and Internet is considered a useful marketing tool to promote heritage tourism in a sustainable manner as it swiftly disseminates information and messages. According to Jang et al., assimilation is the key, for instance, by "integrating multiple subsystems, and connecting themes of different perspectives (including cultural heritage protection, tourism marketing, management and services) to provide a holistic system" (2021, p. 14). Metaverse is becoming the future of digital connectivity.

Metaverse

Buhalis et al. define Metaverse as "the convergence of physical and digital universes, where users can seamlessly traverse between them for working, education and training, health, exploring interests and socialising with others" (2023, p. 3). Ever since Facebook renamed itself as Meta and realigned itself around the Metaverse, the term Metaverse has received unprecedented attention (Buhalis et al., 2023a,b; Glover, 2022; Pew Research Center, 2022). Tourism can benefit immensely from Metaverse (World Economic Forum, 2022) as it offers customization tools. Consumer behavior is anticipated to be different in Metaverse as a broad spectrum of virtual resources can be made accessible (Gursoy et al., 2022). For instance, the consumers can take delight in deeply engaging experiences at the pre-travel stage and access augmented content when they arrive at the destination (Yovcheva et al., 2014). Therefore, it has been predicted that Metaverse holds tremendous potential to heighten the onsite experience of tourists (Buhalis et al., 2022). During the pre-visit stage, they can virtually access various attractions/facilities and determine their appropriateness and appeal and prepare themselves (Dwivedi et al., 2022; Koo et al., 2022). According to Buhalis et al., travelers can immerse themselves to sample a destination or its resource and virtually organize their trip before actual travel (Buhalis et al., 2022). Metaverse tours "are built on existing virtual tours that use 360-degree or 3D media modality technologies. They allow avatar embodiment as well as … physiological immersion through facilitating synchronous interaction with other agents and objects in a virtual destination" (2023, p. 1).

Consequently, Metaverse can shape the manner in which tourism experiences are staged and induce consumer motivations, preferences and onsite behavior. This can lead to innovative marketing and operational designs (Gursoy et al., 2022; Rather, 2023). Buhalis et al. argue that "people who experience destinations and tourism activities virtually tend to be motivated to visit physically. After the visit, Metaverse supports users to engage with places they have already been, share User Generated Content (UGC) in immersive platforms and stay in contact with people they have met, motivating repeat visits and loyalty" (2023a, p. 2). That said, it can also be argued that disruption can happen as travelers will have the flexibility to access virtual resources and digital environments when they are at the destination and this can distract them from enjoying the tangible and intangible attributes of the place (Dwivedi et al., 2022; Koo et al., 2022).

In summary, digital environments can be designed to optimize enjoyment of hyper-realistic virtual interactions and experiences. Undeniably, the critical role of digitalization was evident during the pandemic which led people to completely rely on online networking. The immersive part of online technology is increasingly recognized today for its potential to facilitate spontaneous user interactions in both physical and digital realms. According to Lundmark (2022), "by 2030, every physical device that can be digitally connected will be. Eventually, every action in the digital world will have an effect on the physical world, and vice versa. So, the Metaverse is the coming together of complementary technologies, including cloud and edge (near the data source) computing, artificial intelligence, blockchain, the internet of things, virtual reality, augmented reality and digital twins."

Upon return home, tourists can use Metaverse to recall their experiences, document them or share them, thereby re-living their moments and conversing with others who might have had similar experiences or are seeking for the same. As an example, Buhalis et al. (2023a) referred to a tourist who, after his recent visit to Acropolis in Athens, is able to communicate his experience to other like-minded people. The authors point out that, "Metaverse can lead to memory recalls and increase the intention to return to the destination for both users and also others that experience immersive content. Engagement with tourism material can increase the intention to advocate tourism destinations and organisations as well as to engage with local resources leading to transformative experiences" (p. 5).

Toward a Guided Sustainable Marketing Model

From the foregoing, several new components of an integrated sustainable marketing model for heritage tourism can be identified. A guided sustainable marketing model is proposed in Figure 1.1. It draws on Sheth and Parvatiyar's (2021) dimensional shift model, which calls for a guided visible hand, in the form of policies and political intervention. The underlying purpose is to "actively change markets towards more sustainable behavior, making sustainability goals integral to the business goals" and ensure "corporate marketers embrace transformational strategies to change business practices, particularly those linked to unsustainable production, distribution and consumption" (2021, p. 13). The call is toward more mindful/cognizant and

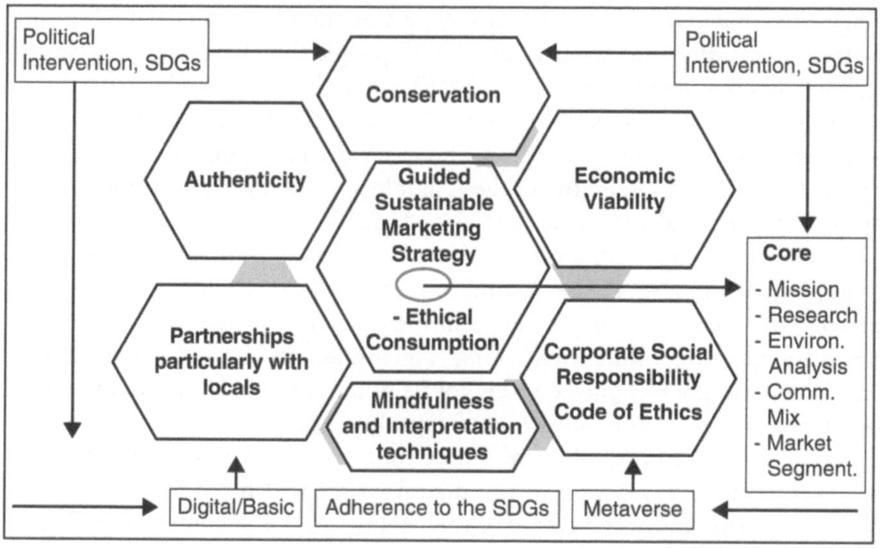

FIGURE 1.1 Guided sustainable marketing model for heritage tourism

meticulous institutional mediation. The model is an extended version of Chhabra's (2010) model. While the inner layer of the model remains the same, it is suggested that it be maneuvered by a guided sustainable marketing strategy promoting ethical consumption in a manner that the sustainable elements in the outer layer are prioritized. Compared to the 2010 version, one additional component is added (social responsibility and code of ethics). It is suggested that sustainable marketing strategy should be guided by SDGs and regulated by political interventions (such as regulatory policies, reform measures, promotional programs and participative partnerships) to ensure adherence to the sustainable marketing agenda promoting authenticity, conservation, partnerships with specific focus on the locals, economic viability and social responsibility and code of ethics. Furthermore, the influence of digital innovations and surging popularity of Metaverse on all role player and functional elements of the marketing model cannot be denied. It is also suggested that the heritage institutions and other stakeholders of heritage tourism should take initiatives to redesign products and services for sustainable use, re-use and post-consumption use, cultivate sustainably conscious behavior in employees as well as the tourists, repurpose marketing mix and reorganize marketing function for sustainability stewardships.

Summary

In summary, the key focus of this chapter was to suggest a guided sustainable marketing model for heritage tourism by advancing the work of existing studies. The next chapter delves deeper into the notion of heritage tourism in the contemporary times. This is followed by a discourse on different transformation theories and their contextualization from a heritage tourism standpoint. A sustainable marketing paradigm for heritage tourism is proposed. Chapter 2 discusses the notions of heritage tourism and transformation in an in-depth manner and suggests a conceptual path toward transformative heritage tourism.

References

Aamir, S., & Atsan, N. (2020). The trend of multisided platforms in the travel industry: Reintermediation of travel agencies and global distribution systems. *Journal of Tourism Futures*, *6*(3), 271–279.

Agyeiwaah, E., Pratt, S., Iaquinto, B. L., & Suntikul, W. (2020). Social identity positively impacts sustainable behaviors of backpackers. *Tourism Geographies*, *25*, 1–22.

Ali, A., Rasoolimanesh, S. M., & Cobanoglu, C. (2020). Technology in tourism and hospitality to achieve sustainable development goals (SDGs). *Journal of Hospitality and Tourism Technology*, *11*(2), 177–181.

Balderjahn, I., Peyer, M., Seegebarth, B., Wiedmann, K. P., & Weber, A. (2018). The many faces of sustainability-conscious consumers: A category-independent typology. *Journal of Business Research*, *91*, 83–93.

Bhattacharya, C. B., & Polman, P. (2017). Sustainability lessons from the front lines. *MIT Sloan Management Review*, *58*(2), 71–78.

Boluk, K., Cavaliere, C., & Higgins-Desbiolles, F. (2017). Call for papers. Critical thinking to realize sustainability in tourism systems: Reflecting on the 2030 sustainable development goals +. *Journal of Sustainable Tourism*, *25*(9), 1201–1204.

Boluk, K., Cavaliere, C. T., & Higgins-Desbiolles, F. (Eds.). (2021). *Activating critical thinking to advance the sustainable development goals in tourism systems*. Routledge.

Boluk, K. A., & Rasoolimanesh, S. M. (2022). Introduction to the special issue on "Deepening our understandings of the roles and responsibilities of the tourism industry towards the United Nations Sustainable Development Goals (SDGs). *Tourism Management Perspectives*, *41*, 100944.

Booms, B. H., & Bitner, M. J. (1981). Marketing strategies and organization structures for service firms. In Donnelly, J. H., and George W. R. (Eds.), Marketing of Services, Conference Proceedings (pp. 47–52). American Marketing Association.

Borland, H., & Lindgreen, A. (2013). Sustainability, epistemology, ecocentric business, and marketing strategy: Ideology, reality, and vision. *Journal of Business Ethics*, *117*, 173–187.

Bramwell, B., Higham, J., Lane, B., & Miller, G. (2017). Twenty-five years of sustainable tourism and the journal of sustainable tourism: Looking back and moving forward. *Journal of Sustainable Tourism*, *25*(1), 1–9.

Buhalis, D., L´opez, E. P., & Martinez-Gonzalez, J. A. (2020). Influence of young consumers' external and internal variables on their e-loyalty to tourism sites. *Journal of Destination Marketing & Management*, *15*, 100409. https://doi.org/10.1016/j.jdmm.2020.100409

Buhalis, D., Lin, M. S., & Leung, D. (2022). Metaverse as a driver for customer experience and value co-creation: Implications for hospitality and tourism management and marketing. *International Journal of Contemporary Hospitality Management*, *35*(2), 701–716. https://doi.org/10.1108/IJCHM-05-2022-0631

Buhalis, D., Leung, D., & Lin, M. (2023a). Metaverse as a disruptive technology revolutionising tourism management and marketing. *Tourism Management*, *97*, 104724.

Buhalis, D., Lin, M. S., & Leung, D. (2023b). Metaverse as a driver for customer experience and value co-creation: Implications for hospitality and tourism management and marketing. *International Journal of Contemporary Hospitality Management*, *35*. https://doi.org/10.1108/IJCHM-05-2022-0631

Cai, Z., Fang, C., Zhang, Q., & Chen, F. (2021). Joint development of cultural heritage protection and tourism: The case of Mount Lushan cultural landscape heritage site. *Heritage Science*, *9*(1), 86.

Chang, H. H., & Sokol, D. D. (2022). How incumbents respond to competition from innovative disruptors in the sharing economy - the impact of Airbnb on hotel performance. *Strategic Management Journal*, *43*(3), 425–446.

Chhabra, D. (2008). Positioning museums on an authenticity continuum. *Annals of Tourism Research*, *35*(2), 427–447.

Chhabra, D. (2009). Proposing a sustainable marketing framework for heritage tourism. *Journal of Sustainable Tourism*. *17*(3), 303–320.

Chhabra, D. (2010). *Sustainable marketing of cultural and heritage tourism*. Routledge.

Chhabra, D. (2012). A present-centered dissonant heritage management model. *Annals of Tourism Research*, *39*(3), 1701–1705.

Chhabra, D. (2015). *Strategic marketing in hospitality and tourism: Building a 'SMART' online agenda*. Nova Science Publishers, Inc.

Chhabra, D. (2021). *Resilience, authenticity and digital heritage tourism*. Routledge.

Chi, C. G. Q., Zhang, C., & Liu, Y. (2019). Determinants of corporate social responsibility (CSR) attitudes: Perspective of travel and tourism managers at world heritage sites. *International Journal of Contemporary Hospitality Management*, *31*(6), 2253–2269.

Cravens, D. W. (1974). Marketing management in an era of shortages. *Business Horizons*, *17*(1), 79–85.

Cuculeski, N., Petrovska, I., & Cuculeski, V. (2016). Sustainable marketing and consumers' preferences in tourism. *European Journal of Tourism, Hospitality and Recreation*, *7*(2), 84–90.

Du Cros, H. (2001). A new model to assist in planning for sustainable cultural heritage tourism. *International Journal of Tourism Research*, *3*(2), 165–170.

Dwivedi, Y. K., Hughes, L., Baabdullah, A. M., Ribeiro-Navarrete, S., Giannakis, M., Al-Debei, M., Dennehy, D., Metri, B., Buhalis, D., Cheung, C., Conboy, K., Doyle, R., Dubey, R., Dutot, V., Felix, R., Goyal, D., Gustafsson, A., Hinsch, C., Jebabli, I. et al (2022). Metaverse beyond the hype: Multidisciplinary perspectives on emerging challenges, opportunities, and agenda for research, practice and policy. *International Journal of Information Management*, 66, 102542.

Ferrell, L., & Ferrell, O. C. (2022). Broadening the definition of socially responsible marketing. *Journal of Macromarketing*, *42*(4). https://doi.org/10.1177/02761467221094570

Fisk, G. (1973). "Criteria for a theory of responsible consumption. *Journal of Marketing*, *37*(2), 24–31.

Glover, G. (2022). Bank of America shares how to invest in the metaverse and 13 other disruptive technologies that could transform our lives and become a market worth over $6 trillion by 2030. February. Retrieved from https://www.businessinsider.com/investing-strategy-tech-6g-metaverse-flying-cars-ai-moonshots-2021-9

Gursoy, D., Malodia, S., & Dhir, A. (2022). The metaverse in the hospitality and tourism industry: An overview of current trends and future research directions. *Journal of Hospitality Marketing & Management*, *31*(5), 527–534.

Guttentag, D. A., & Smith, S. L. J. (2017). Assessing Airbnb as a disruptive innovation relative to hotels: Substitution and comparative performance expectations. *International Journal of Hospitality Management*, *61*, 1–10.

Hall, C. M. (2019). Constructing sustainable tourism development: The 2030 agenda and the managerial ecology of sustainable tourism. *Journal of Sustainable Tourism*, *27*(7), 1044–1060.

Henion, K. E. (1976). *Ecological marketing*. Grid.

Horng, J. S., Hsu, H., & Tsai, C. Y. (2018). An assessment model of corporate social responsibility practice in the tourism industry. *Journal of Sustainable Tourism*, *26*(7), 1085–1104.

Hult, G. T. M., Mena, J. A., Gonzalez-Perez, M. A., Lagerström, K., & Hult, D. T. (2018). A ten country-company study of sustainability and product-market performance: Influences of doing good, warm glow, and price fairness. *Journal of Macromarketing*, *38*(3), 242–261.

Hunt, S. D. (2011). Sustainable marketing, equity, and economic growth: A resource-advantage, economic freedom approach. *Journal of the Academy of Marketing Science*, *39*(1), 7–20.

Jang, S., Chong, K., & Yoo, C. (2021). The effect of mobile application-driven customer participation on bakery purchase behavior: Evidence from a field experiment. *International Journal of Hospitality Management*, *94*, 102865.

Kolar, T., & Koklič, M. K. (2018). Managing sustainable consumption of cultural heritage: The key role of existential authenticity. In A. Campelo, L. Reynolds, A. Lindgreen, & M. Beverland (Eds.), *Cultural heritage* (pp. 71–84). Routledge.

Koo, C., Kwon, J., Chung, N., & Kim, J. (2022). Metaverse tourism: Conceptual framework and research propositions. *Current Issues in Tourism*. https://doi.org/10.1080/13683500.2022.2122781

Kotler, P., Bowen, J., Makens, J., & Baloglu, S. (2016). *Marketing for hospitality and tourism* (7th ed.). Pearson.

Kramer, M. R. (2020), Larry Fink isn't going to read your sustainability report. Harvard Business Review (January 20). HBR.ORG.

Laczniak, G. R., & Murphy, P. E. (2018). The role of normative marking ethics. *Journal of Business Research*, *95*, 401–407.

Laczniak, G. R., & Santos, N. J. (2011). The integrative justice model for marketing to the poor: An extension of SD logic to distributive justice and macromarketing. *Journal of Macromarketing*, *31*(2), 135–147.

Laczniak, G., & Shultz, C. (2021). Toward a doctrine of socially responsible marketing (SRM): A macro and normative-ethical perspective. *Journal of Macromarketing*, *41*(2), 201–231.

Laczniak, G., & Shultz, C. (2022). Socially responsible marketing (SRM): Broader, deeper, and ever greater. *Journal of Macromarketing*, *42*(4), 554–559.

Lin, C. P., & Ding, C. G. (2005). Opening the black box: Assessing the mediating mechanism of relationship quality and the moderating effects of prior experience in ISP service. *International Journal of Service Industry Management*, *16*(1), 55–80.

Loureiro, S. M. C., Guerreiro, J., & Ali, F. (2020). 20 years of research on virtual reality and augmented reality in tourism context: A text-mining approach. *Tourism Management*, *77*, 104028.

Luan, C. C., & Chen, T. H. (2021). Empathy from private or public self-consciousness in socially responsible consumption. *Journal of Philanthropy and Marketing*, *26*(3), e1695.

Lundmark, P. (2022). *The real future of the metaverse is not for consumers*. Financial Times.

McCarthy, E. J. (1960). *Basic marketing: A managerial approach*. Irwin.

McClements, D. J., Barrangou, R., Hill, C., Kokini, J. L., Lila, M. A., Meyer, A. S., & Yu, L. (2021). Building a resilient, sustainable, and healthier food supply through innovation and technology. *Annual Review of Food Science and Technology*, *12*, 1–28.

McLean, F. (2002). *Marketing the museum*. Routledge.

Meadows, D. H., Meadows, D. L., Randers, J., & Behrens, W. H. III (1972). *The limits to growth: A report for the club of Rome's project on the predicament of mankind*. Universe Books.

Misiura, S. (2006). *Heritage marketing*. Elsevier.

Mitchell, R. W., Wooliscroft, B., & Higham, J. (2010). Sustainable market orientation: A new approach to managing marketing strategy. *Journal of Macromarketing*, *30*(2), 160–170.

Mittelstaedt, J., Shultz, C. J., Kilbourne, W. E., & Peterson, M. (2014). Sustainability as megatrend: Two schools of marketing thought. *Journal of Macromarketing*, *34*(3), 253–264.

Morrison, A. (2010), *Hospitality and travel marketing* (4th ed.). Delmar Cengage Learning.

Mousazadeh, H., Ghorbani, A., Azadi, H., Almani, F. A., Zangiabadi, A., Zhu, K., & Dávid, L. D. (2023). Developing sustainable behaviors for underground heritage tourism management: The case of Persian Qanats, a UNESCO world heritage property. *Land*, *12*(4), 808.

Nill, Alexander and Bianka Papp (2020). Corporate social responsibility in the United States: A historic perspective on the interplay between ideologic, political and economic forces. In Schlegelmilch Bodo (Ed.), *Rethinking business responsibility in a global context* (pp. 75–100). SpringerLink.

Nowacki, M., & Kruczek, Z. (2021). Experience marketing at Polish museums and visitor attractions: The co-creation of visitor experiences, emotions and satisfaction. *Museum Management and Curatorship*, *36*(1), 62–81.

Olya, H. G., Alipour, H., & Gavilyan, Y. (2018). Different voices from community groups to support sustainable tourism development at Iranian World Heritage Sites: Evidence from Bisotun. *Journal of Sustainable Tourism*, *26*(10), 1728–1748.

Passafaro, P. (2020). Attitudes and tourists' sustainable behavior: An overview of the literature and discussion of some theoretical and methodological issues. *Journal of Travel Research*, *59*, 579–601.

Perry, D. L. (1976). *Social marketing strategies. Conservation issues and analysis*. Goodyear Publishing.

Pew Research Center. (2022). *The metaverse in 2040*. Pew Research Center.

Polonsky, M. J., & Rosenberger, P. J. III (2001). Reevaluating green marketing: A strategic approach. *Business Horizons*, *44*(5), 21–30.

Pomering, A., Noble, G., & Johnson, L. W. (2011). Conceptualising a contemporary marketing mix for sustainable tourism. *Journal of Sustainable Tourism*, *19*(8), 953–969.

Prothero, A., & McDonagh, P. (2015). Introduction to the special issue: Sustainability as a megatrend II." *Journal of Macromarketing*, *35*(1), 7–10.

Rasoolimanesh, S. M., Ramakrishna, S., Hall, C. M., Esfandiar, K., & Seyfi, S. (2020). A systematic scoping review of sustainable tourism indicators in relation to the sustainable development goals. *Journal of Sustainable Tourism*, *31*(7), 1497–1517.

Rather, R. A. (2023). Metaverse marketing and consumer research: Theoretical framework and future research agenda in tourism and hospitality industry. *Tourism Recreation Research*, 1–9. https://doi.org/10.1080/02508281.2023.2216525.

Rauschnabel, P. A., Felix, R., Hinsch, C., Shahab, H., & Alt, F. (2022). What is XR? Towards a framework for augmented and virtual reality. *Computers in Human Behavior*, *133*, 107289.

Rowan, Y., & Baram, U. (Eds.). (2004). *Marketing heritage. Archaeology and the consumption of past*. Altamira Press.

Santos, N. J. C., & Laczniak, G. R. (2009). Marketing to the poor: An integrative justice model for engaging impoverished market segments. *Journal of Public Policy and Marketing*, *28*(1), 3–15.

Scheyvens, R., & Cheer, J. M. (2021). Tourism, the SDGs and partnerships. *Journal of Sustainable Development*. https://doi.org/10.1080/09669582.2021.1982953

Seyfi, S., Hall, C. M., & Rasoolimanesh, S. M. (2020). Exploring memorable cultural tourism experiences. *Journal of Heritage Tourism*, *15*(3), 341–357.

Shapiro, S. (2022). Building on a long needed seminal contribution. *Journal of Macromarketing*, *42*(4). https://doi.org/10.1177/02761467221087356

Shapiro, S. J. (1978). Marketing in a conserver society. *Business Horizons*, *21*(2), 3–13.

Sheth, J., Sethia, N., & Srinivas, S. (2011). Mindful consumption: A customer-centric approach to sustainability. *Journal of the Academy of Marketing Science*, *39*(1), 21–29.

Sheth, J. N., & Parvatiyar, A. (2021). Sustainable marketing: Market-driving, not market-driven. *Journal of Macromarketing*, *41*(1), 150–165.

Sheth, J. N., & Parvatiyar, A. (2022). Socially responsible marketing: Toward aligning Dharma (Duties), Karma (Actions), and Eudaimonia (Well-being). *Journal of Macromarketing*, *42*(4), 590–602.

Sheth, J. N., Sethia, N. K., & Srinivas, S. (2011). Mindful consumption: A customer-centric approach to sustainability. *Journal of the Academy of Marketing Science*, *39*(1), 21–29.

Sheth, J. N., & Wright, P. L. (1974). *Marketing analysis for societal problems*. Bureau of Economic Research, College of Commerce and Business Administration, University of Illinois.

Shobeiri, S., Rajaobelina, L., Durif, F., & Boivin, C. (2016). Experiential motivations of socially responsible consumption. *International Journal of Market Research*, *58*(1), 119–139.

Shultz, C. J., & Holbrook, M. B. (1999). Marketing and the tragedy of the commons: A synthesis, commentary, and analysis for action. *Journal of Public Policy and Marketing*, *18*(2), 218–29.

Stevens, P. (2019), Behind Walmart's Push to Eliminate 1 Gigaton of Greenhouse Gases by 2030, CNBC, Dec. 15. Accessed March 15, 2020. https://www.cnbc.com/2019/12/15/walmarts-projectgigaton-is-its-most-ambitious-climate-goal-yet.html

Tandrayen-Ragoobur, V., Tengur, N. D., & Fauzel, S. (2022). COVID-19 and Mauritius' tourism industry: An island perspective. *Journal of Policy Research in Tourism, Leisure and Events*.

Tahiri, A., Kovaçi, I., & Trajkovska Petkoska, A. (2022). Sustainable tourism as a potential for promotion of regional heritage, local food, traditions, and diversity—case of Kosovo. *Sustainability*, *14*(19), 12326.

Timothy, D. J. (2021). *Cultural heritage and tourism: An introduction* (2nd ed.). Channel View Publications.

Tleuberdinova, A., Shayekina, Z., Salauatova, D., & Pratt, S. (2021). Macroeconomic factors influencing tourism entrepreneurship: The case of Kazakhstan. *The Journal of Entrepreneurship*, *30*(1), 179–209.

Ulusoy, E. (2016). Experiential responsible consumption. *Journal of Business Research*, *69*(1), 284–297.

United Nations. (2023). The Sustainable Development Goals Report 2023: Special Edition. Retrieved from https://unstats.un.org/sdgs/report/2023/

UNWTO. (2018). Tourism and the sustainable development goals: Journey to 2030. https://www.e-unwto.org/doi/pdf/10.18111/9789284419401

UNWTO & UNDP. (2017). *Tourism and the sustainable development goals – Journey to 2030*. UNWTO.

van Dam, Y. K., & Apeldoorn, P. A. C. (1996). Sustainable marketing. *Journal of Macromarketing*, *16*(2), 45–56.

Varadarajan, R. (2017). Innovating for sustainability: A framework for sustainable innovations and a model of sustainable innovation orientation. *Journal of the Academy of Marketing Science*, *45*(1), 14–36.

Varadarajan, R. (2019). Theoretical underpinnings of research in strategic marketing: A commentary. *Journal of the Academy of Marketing Science*, *47*, 30–36.

Varey, R. J. (2010). Marketing means and ends for a sustainable society: A welfare agenda for transformative change. *Journal of Macromarketing*, *30*(2), 112–126.

Varey, R. J. (2012). The marketing future beyond the limits of growth. *Journal of Macromarketing*, *32*(4), 421–430.

Vila, O. R., & Bharadwaj, S. (2019). Pioneering research on sustainability. In A. Parvatiyar, & R. Sisodia (Eds.), *Handbook of advances in marketing in an era of disruptions – Essays in honour of Jagdish N. Sheth* (pp. 125–132). Sage.

Villena, V. H., & Gioia, D. A. (2020), A more sustainable supply chain. *Harvard Business Review* (March–April), HBR.ORG.

Weiermair, K., Peters, M., & Schuckert, M. (2007). Destination development and the tourist life-cycle: Implications for entrepreneurship in Alpine tourism. *Tourism Recreation Research*, *32*(1), 83–93.

Whelan, T., & Fink, C. (2016), The comprehensive business case for sustainability. *Harvard Business Review*, October 21. HBR.ORG

Wilkie, W. L., & Moore, E. S. (1999). "Marketing's contributions to society. *Journal of Marketing*, *63*(4), 198–218.

Winston, A. (2019), "The top sustainability stories of 2019," Harvard Business Review, December 30. HBR.ORG.

World Commission on Environment and Development (WCED). (1987). *Our common future*. Oxford University Press.

World Economic Forum. (2022). How the metaverse can be a force for good in an uncertain world. Retrieved from https://www.weforum.org/agenda/2022/05/how-metaverse-can-be-a-force-for-good-in-an-uncertain-world

Xie, P. (2024). Heritage Tourism as Transformation in Staycation Era. In *D. Chhabra (Ed.), Sustainable marketing of transformative heritage tourism*. Routledge. In Press.

Yovcheva, Z., Buhalis, D., Gatzidis, C., & van Elzakker, C. P. (2014). Empirical evaluation of smartphone augmented reality browsers in an urban tourism destination context. *International Journal of Mobile Human Computer Interaction (IJMHCI)*, 6(2), 10–31. https://doi.org/10.4018/ijmhci.2014040102

2

CONCEPTUALIZING TRANSFORMATIVE HERITAGE TOURISM

Deepak Chhabra

Heritage tourism, an alternative form of tourism, has become a catchword for several decades. Heritage is a nebulous concept and therefore subject to myriad interpretations and expressions. It has been examined extensively from commodification and conservation standpoints. Furthermore, it has been treated discursively from both positive and negative viewpoints. As encapsulated by Merriman, the favorable side of heritage strives to safeguard culture and its various manifestations for sustained use, whereas its negative aspect refers to "manipulation and exploitation of the past for commercial ends" (1991, p. 8). Heritage can be broadly classified into tangible and intangible heritage that exist in rural, urban and suburban settings. Prentice (1993) has categorized heritage into three types: historic and artistic (with tangible traits); scientific (associated with features sketched from animals, birds, rocks, trees, etc.) and cultural heritage (such as folklife, languages and traditions). From a tourism standpoint, heritage refers to cultural and natural features of a landscape. It "represents some sort of inheritance to be passed down to current and future generations, both in terms of cultural traditions and physical artifacts" (Timothy & Boyd, 2003, p. 6). McLean (2002) writes that the conjunction between tourism and heritage applications illustrates how power can shape the manner heritage is showcased for tourism purpose. Chhabra points out that "heritage is a driving force of complexes which focuses on disseminating cultural capital. Thus, within the tourism context, heritage has become a commodity aimed to fulfill the needs of the contemporary tourist" (2010, p. 2). Taylor (2001) contends that monetary remunerations constitute the driving force behind the promotion of heritage tourism. The economic goals and significance stimulate the need to construct a common dais to match heritage supply with heritage demand. The economic

DOI: 10.4324/9781003346197-2

significance of marrying heritage with tourism cannot be underestimated (Lak et al., 2020; Mandić & Kennell, 2021; Snowball, 2013; Timothy, 2018, 2020). In fact, extant literature has focused on economic benefits of heritage tourism (Adie & Falk, 2021; Chhabra et al., 2003; Dans & González, 2018; Petronela, 2016).

Defining Heritage Tourism

Evidently, different definitions of heritage tourism exist. These can be broadly delineated based on attributes such as motivations, attributes/resources, past-connectedness and nostalgia/emotions and experiences. For instance, Timothy defines heritage tourism as traveling to see or experience "built heritage, living culture or contemporary arts" (2021, p. 4). According to him, heritage tourism visits are "motivated by a desire to enhance one's one cultural self, to learn something new, to spend time with friends and family, to satisfy one's own curiosity or simply to use up excess time. In short, heritage tourism encompasses a multitude of motives, resources and experiences" (Timothy, 2021, p. 4). In a nutshell, answer is complex to the question: what is heritage tourism? According to Halewood and Hannam, a broad spectrum of themes forms the underpinnings of various definitions of heritage tourism; for instance, themes include "analysis of museums, landscapes, artifacts, and activities that concentrate on representing different aspects of the past" (2001, p. 566). Some definitions are polarized and grounded in demand- and/or supply side overtones.

Moscardo (2001) refers to heritage tourism as an experience stimulated by interactions between the visitors and the heritage resources. Poria et al. (2001) describe heritage tourism as a phenomenon emanated by perceptions of visitors about a site. Cognitive perceptions, expectations and motivations are often used to coin demand-side definitions. Examples of heritage tourism motivations include nostalgia, self-esteem and quest for an 'authentic' experience. As aptly described by Lowenthal, "if the past is a foreign country, nostalgia has made it the foreign country with the healthiest tourist trade of all" (1985, p. 4). In other words, marketing messages can appeal to nostalgic emotions.

Chhabra proposed several fundamental tenets of heritage tourism such as "economics, emotions/motivation, inheritance, past, authenticity and participation" (2010, p. 4). Some scholars have envisioned that the key intention of heritage tourism should be to build a marketable and palatable product for tourists, in a strategically planned and structured/organized manner. Many studies have endeavored to obtain insights on what makes a heritage tourism experience unique and note that its uniqueness can be attributed to history, authentic reenactment of events from the bygone era or drawing on the authentic icons of the past (Chhabra et al., 2003; Chung et al., 2018;

Yi et al., 2022). Heritage tourism, being past-oriented, requires chronological distance between the historical or cultural past of, for instance, the events and its present expressions (Leighton, 2007). In other words, heritage tourism aims at creating "a temporal distance between the tourist and the place" (Scarpi & Raggiotto, 2023, p. 1). Recent studies have drawn attention to the wellness potential of heritage tourism. Research scrutinizing this line of inquiry is still in its nascent stages. Brooks et al. state that "heritage tourism fosters host communities' participation and enhances their capacity to flourish and support long-term health and wellbeing" (2023, p. 1). The authors examine the impact of heritage tourism on sustainable community development and the health and well-being of local communities. Similar results are supported by Huang et al. (2024).

Therefore, it is important to acknowledge the far-reaching inferences of heritage tourism (Apostolakis, 2003). For the purpose of this study, heritage tourism is defined as follows:

> the phenomenon that draws on the chronological distance between the legacy of yesteryears and the present. It is about the management of the past, inheritance, and authenticity to heighten interest, participation and gratification consumer yearning for the bygone era by evoking nostalgic emotions and memories with an underlying purpose of stimulating economic benefits and wellbeing for its various constituencies.

Heritage Tourism Trends

Marketing managers should take emerging trends into consideration as they can impact "the planning and success thereof the marketing strategies with the realm of heritage tourism" (Chhabra, 2010, p. 14). While some heritage tourism trends continue over the past decade, new trends have emerged. These include the following:

- *Authenticity*: It remains the cornerstone of heritage tourism and its scholars continue to inspect it from numerous perspectives (DeSoucey et al., 2019; Halewood & Hannam, 2001; Moore et al., 2021). Insights on recent discourses of authenticity are shared in a following section.
- *Wellness*: Recent studies have become attentive to wellness-related benefits of heritage tourism (Brooks et al., 2023; Heritage Tourism Outlook, 2023).
- *Digitalization*: The heritage tourism industry today acknowledges the critical role of technology to promote sustainable and immersive experiences in addition to facilitating collaboration between key stakeholders. In fact, UNESCO has prioritized the digital transformation of heritage conservation (Zhang & Zou, 2022). Digitization of heritage can aid in modernizing heritage tourism to a certain extent (Zhenrao et al., 2021).

- *Psychological distance*: It is noted that heritage tourism induces tourists to embrace a higher psychological distance and therefore it increases the construal level (Trope & Liberman, 2010). Sequentially, "this higher construal level negatively affects destination loyalty and perceived uniqueness. However, authenticity and engagement moderate the heritage–construal relationship, counterbalancing the higher psychological distance induced by heritage" (Liu et al., 2023, p. 3).
- *Integration with the SDGs (Sustainable Development Goals)*: There is an emerging need to align heritage conservation goals with strategies that promote sustainable cultural heritage tourism in a manner that there is harmony and shared sense of purpose among the key stakeholders. Specifically, SDG 11 and SDG 17 need to be targeted to minimize externalities (Liu et al., 2023).
- *Barriers to SDGS*: Heritage sites are presently under threat from a variety of factors, including degradation of the built fabric (Sevieri & Galasso, 2021), development disputes (Xiao et al., 2018), tourism development (Ensenat-Soberanis & Blanco-Gregory, 2022) and natural catastrophes (Osello et al., 2018), all of which impede the accomplishment of sustainable development goals (Xiao et al., 2018), particularly sustainable tourism.
- *Insights on stakeholders*: The stakeholders are the most essential strand in the development of sustainable heritage tourism (Chhabra, 2010, 2021; Elabd et al., 2021). There is on ongoing need to be better informed of the aims, interests and values of important partners to improve collaboration and collectively accomplish sustainable objectives (Alazaizeh et al., 2020).
- *Cultural policies*: Cultural policies implemented by government sectors are critical for equitable development, since they can regulate not only equal cost–benefit distributions, but also inclusive planning and decision-making practices (Rastegar et al., 2021).
- *History travel*: It is an emerging sub-set of heritage tourism. For instance, the historical trips led by historians and experts are approved by the History Channel and they provide on-the-spot, topic-specific information and historical commentary of the visited destination. The expertise of the historians can provide a crucial context and connect the individual experience of each traveler to the larger historical picture.
- *Surge in the travelers seeking packaged trips*: With numerous companies offering distinct services and all-inclusive programs, the 'Packaged Traveler' has become the majority class within the heritage tourist segment. Heritage tourism programs can be offered that feature a variety of activities, locations and price ranges, thereby meeting the needs of the majority of the market without requiring them to search several sources for various services. Packaged trips can aid in saving time and money.

- Religion-*based travels*: Religion-based travels also play an important role in promoting heritage tours and travels. Every year, millions of religious travelers travel from one part of the world to the other.
- *Potential to boost social and economic development*: Heritage, culture and nature have become significant tools for social and economic development of host regions. Countries can leverage their rich history and heritage to enhance their brand image (Chhabra, 2021; Heritage Outlook, 2023; Sands, 2023; Timothy, 2021).

Authenticity

As mentioned earlier, authenticity continues to be an enigmatic and multifaceted concept and its dynamic nature has been scrutinized extensively in literature from numerous standpoints. A plethora of studies continue to refer to actual, unaltered, progressive and constancy versions of authenticity as a point of reference. It is noted that in some cases, tourists enjoy the pseudo/counterfeit heritage experiences, whereas, in other scenarios, tourists quest for genuinely and objectively authentic experiences endorsed by legitimized sources (Timothy, 2021). In recent times, a dialogical view of authenticity continues to oscillate although the theories that elucidate the authenticity geniuses remain exuberant and continue to enrich its dialectical nature from multifaceted perspectives (Moore et al., 2021). Capturing the essence of various perspectives of authenticity, authenticity can be broadly delineated into five categories: objective/essentialist, constructivist, negotiated, existentialist and theoplacity (Belhassen et al., 2008; Chhabra, 2021; Rickly, 2022; Timothy, 2021).

Essentialist or objective version is considered a proxy to the actual, original and unaltered account of authenticity (Chhabra, 2010; Timothy, 2021; Wang, 1999). Constructivist theory of authenticity refers to commodified versions of authenticity to appeal to market demand and examples in this category include pseudo, purposely staged and contrived settings. The negotiated version of authenticity is positioned as a midpoint on the essentialist/commodification continuum. This stance is "slanted to rationalize a midway point, by retaining originality or embracing sanitized modifications, to meet consumer demand" (Chhabra, 2022, p. 3). This position is co-created between the consumers and the suppliers of heritage tourism (Adams, 1996). According to DeSoucey et al. (2019), the negotiated version assists in protecting important aspects of heritage and fortifying the core features of objective authenticity. Wang (1999) connotes the existentialist view to optimized 'truth of the moment' experiences which stimulate connection with one's true self. Because it is focused on experiential consumption, it is subjective and refers to the state of mind and holds potential to generate eudaimonic benefits (Gino et al., 2015; Kirillova & Lehto, 2015;

Mkono, 2020). The theoplacity view of authenticity positions "tourists in environmental and sociopolitical contexts; it recognizes the complex notion of authenticity shaped by interactive dialogue between the visited place, trust, encounter, and the self" (Belhassen et al., 2008, p. 685). Confirming this view, recent literature notes that "a tourist enjoys an optimized experience, feels a sense of happiness, and is in touch with himself or herself in an essentialist (original/genuine) setting" (Chhabra, 2021, p. 5). Some scholars use different terms, such as iconic and referential authenticity, for these delineations but denote similar meanings.

Despite the fact that extant literature focuses on the authenticity phenomenon, it is interesting to note that authenticity in heritage tourism has rarely been discussed from a marketing standpoint, particularly in the context of sustainability. From a marketing viewpoint, consumers quest for authenticity has become a linchpin of contemporary marketing agenda (Timothy, 2021). For instance, some scholars have reported significant positive relationship between authenticity and the desire to consume cultural heritage (DeSoucey et al., 2019; Rickly, 2022; Timothy, 2021). As the online world booms and becomes more boisterous and crowded, marketing strategies centered on authenticity can offer a competitive edge. An authentic disposition in marketing guides intentional sincerity/truthfulness toward the product/service, for instance, by designing appropriate messages. Chhabra and Ashton (2023) note that "consumers are lured by enterprises that truthfully present themselves, share their core values, and demonstrate dedicated support for a social cause. Authenticity, from a marketing standpoint, is mushrooming as an important field of inquiry in tourism." In this context, it extends beyond promotional campaigns and touches every aspect of an enterprise and/or corporation. However, authenticity is not clearly understood in its market expressions. Therefore, there is a need for a shift in marketing paradigms to ensure that authenticity is centered at the core of all promotional mix strategies. In fact, one touted pathway is a transformative approach that holds potential to facilitate the use of authenticity in promotional strategies in a socially responsible/ethical/sustainable manner. The significant association between authenticity and transformation and well-being is reported by only a handful of studies (Chhabra, 2021; Kirillova et al., 2017a, 2017b; Pung et al., 2020).

Transformation

This section offers insights on the notion of transformation, its important theories and triggers. Notable transformation paradigms are comprehended. Deliberations are also offered on the manner in which authenticity is contextualized. Intrinsic pathways and transformative spaces are touched upon.

Conceptualizing Transformation

Transformation in documented literature is described as a "positive change in attitude, behavior, and/or personality" (Pritchard & Morgan, 2013, p. 2). Transformation is the eventual purpose in tourist ventures. According to Pritchard and Morgan, "human survival is inextricably linked to human transformation" (2013, p. 3). It is important to understand the phenomenon of transformation in its entirety, understand the processes and factors that trigger transformation and identify its various dimensions and features. Most definitions of transformation in their nascent stages have looked at this phenomenon from an adult learning standpoint. For instance, according to O'Sullivan, transformative learning is "a shift of consciousness that dramatically and irreversibly alters our way of being in the world" (2002, p. 11), thereby referring to a shift in consciousness which results in a different or new way of being. Yang et al. refer to transformation as "... positive changes leading to a natural status free of physical or mental diseases" (2015, p. 1884). Collins-Kreiner connects transformation with "... enlightenment, life-changing events, and consciousness-changing events" (2010, p. 445). Other definitions signify cognition and affective as dimensions of transformation while noting that identity change is the ultimate outcome of transformation (Ulusoy, 2016). Focusing on the cognitive aspect of transformation, Kirillova et al. define transformation as "... a growth enhancing, irreversible change that is a fundamental break with the past or current practices that requires new knowledge for successful implementation" (2017b, p. 501). Coghlan and Weiler define transformation as "... an individualized process which can lead to a critical awareness of the self, leading to a new self-definition" (2018, p. 567). This definition mirrors the transformation process and its cognitive features such as mindfulness and a fresh understanding of the self. Recurring themes from the aforementioned definitions can be grouped into cognition, affective and conation categories. In fact, these are captured in the Tri-Component Attitude (TCA) Model (Agyeiwaah et al., 2021; Rosenberg & Hovland, 1960). By referring to mindfulness and deep change in multiple dimensions related to a person such as self, cognitive, emotional, unconscious and somatic, Mezirow stresses on the cognitive and emotional features of the TCA model in addition to the self and its unconscious aspects. Taking the notion of transformative learning to the next level, O'Sullivan (2002) refers to a change in consciousness of the self as well as the surrounding environment resulting in an increased level of mindfulness and responsibility toward the negative impacts and the marginal populations. The affective dimension has been deemed to be more significant by some scholars as "extraordinary experiences create arousal of emotions dominating the cognitive dimension" (Carù & Cova, 2003; Lindberg & Østergaard, 2015).

Studies focusing on tourism have reported congruent insights on cognitive, affective and conative domains. As an instance, Barbieri and Marmol (2022) note transformation in volunteer tourists such as increased levels of social responsibility and allegiance to reduce poverty. Kirillova et al. (2017b) write that a "transformative experience is an extended and enhanced version of a 'peak' experience" (2017b, p. 508). In congruence, Ulusoy (2016) identify three facets of transformative experience (TE): enhanced empowerment, allegiance to commitment to accountability and mindful behavior. Pung et al. (2020) note transformative changes in a person in the realm of skills, cognizance, daily routine and behavior. Transformations have also resulted in new social identities, interpersonal authenticity and solidarity (Everett & John Parakoottathil, 2018; Soulard et al., 2021).

Niche forms of tourism, such as volunteer tourism, backpacker tourism, educational tourism and heritage tourism, hold potential to inspire/trigger more profound changes. For instance, Brown (2013) has studied TEs from the perspective of reinstating a state of existential authenticity in the self by referring to alteration in tourist values and behavior through more immersive engagements over an extended period of time. Values and higher-level needs such as self-fulfillment and an elevated sense of consciousness are also noted by several scholars. Lean (2009) states that lasting changes in "values, knowledge, attitudes, and behavior which contributes to not only an individual's wellness but also that of the greater good. Self-awareness, self-renewal, ..., heightened levels of independence and empowerment are considered to be some of the outcomes of volunteerism" (Tasci & Godovykh, 2021, p. 3).

It is also necessary to cognize that transformation is a process than spans several stages (Chhabra, 2022; Coghlan & Weiler, 2018; Mezirow, 1991; Pung et al., 2020; Sheldon, 2020). As an instance, Pung et al. (2020) detect various transformation stages beginning with subtle, dilemmas and cultural alienation, resulting in optimal experiences, contemplation, assimilation and alteration in attitude, traditions and behavior. The authors define the transformation as a process that "is facilitated by contextual stimuli which strike the tourists and lead to reflecting and integrating new knowledge, skills and beliefs, which ultimately enhance the tourists' existential authenticity and increase the tourists' cross-cultural understanding and pro-environmental awareness, with potential consequences on long-term behavior" (Pung et al., 2020, p. 2). At this juncture, transformation is described from both antecedent and consequential standpoints.

Transformative Theory

Several scholars have anchored their transformation discourses on Mezirow's theory of transformative learning. Mezirow's theory suggests a ten-step transformation process and postulates that an individual becomes cognizant of his or her positionality upon recognizing the existence of larger environment factors which are beyond the person's control. These include political, economic, sociocultural and spiritual forces. Mezirow's (1991) proposed theory of transformative learning consists of ten steps, to acquire a transformed state of mind:

1 experience a disorienting dilemma;
2 undergo self-examination;
3 conduct a deep assessment of personal role assumptions and alienation created by new roles;
4 share and analyze personal discontent and similar experiences with others;
5 explore options for new ways of acting;
6 build competence and self-confidence in new roles;
7 plan a course of action;
8 acquire knowledge and skills for action;
9 try new roles and assess feedback and
10 reintegrate into society with a new perspective.

Coughlan and Weiler (2018) employ Mezirow's theory in the context of volunteer tourism and identify five steps: (1) the antecedents and setting for transformative learning to happen (i.e. looking for immersive engagement with the social and natural realms and an alteration in the situation that triggers opportunities for transformative learning), (2) bewildering dilemmas, and the surfacing of perplexing emotions (associated with the first three steps of Mezirow's theory), (3) the significance of dialect and contemplation (Steps 4 and 5), (4) self-fulfillment as a consequence of volunteer tourism (Steps 6 and 8) and finally (5) integrating back in society (Step 10). The authors add two more themes to Mezirow's theory. The first pertains to disenchantment and the quest for new points of reference to stimulate desire for participation, for instance, in a volunteer tourism project. The second theme refers to the need for creating a situation or setting that is conducive toward the progression of different phases. As mentioned earlier, several scholars have noted that alternative forms of tourism (such as volunteer tourism, backpacker tourism, wellness tourism and spiritual tourism) provide an ideal 'othered' setting and conducive environment to trigger transformation.

Coghlan and Gooch further note that "engagement in transformative learning appears to be driven by an albeit subconscious disillusion with existing sets of reference and the desire to seek something novel" (2011, p. 719). According to Lange (2004), people in Western countries have a tendency to encounter a sense of alienation as they feel constrained in their daily routine. Taylor (2007) points out that critical self-reflexivity can be triggered when one distances oneself from routine life and seeks to experience new learning prospects. Yorks and Kasl (2006) focus on holistic learning environments for transformative learning. The authors share that experiential learning tools can facilitate deeply felt experiences and intuitive expressions/deliberations that are rooted in emotions and new knowledge drawn from narrative and communal sagas, thereby building a lived experience for the person to deliberate on. O'Sullivan (2002) stresses on the significance of inculcating a sense of place through education of community history and of inhabitants of a bioregion and argue that this approach can assist in overcoming a sense of dislocation.

Transformative Experiences

People quest for self-actualization, an elevated state of mind. As volunteers traverse through Mezirow's ten phases, they make an effort to reassess their values and develop new capabilities and self-confidence through new aptitude and learning. Other related topics are authenticity, the personal self (we become aware of our distinct thought processes/way of thinking in due course of our life time and the manner in which we portray our novel real self in the society) and emancipatory learning (emphasizing on practice and social change outcomes) (Cranton & King, 2003). Many scholars use the examples of international travel and volunteer activities to illustrate how they are motivated by a desire for personal growth and the need for acquiring new knowledge enhance the self (Wearing & Neil, 2001). Within these contexts, TEs are recognized for producing benefits in terms of (1) personal development and self-fulfillment, particularly a great sense of empowerment and independence, increasing self-awareness and a spiritual development by reflecting on Western values (Brown & Lehto, 2005; Wearing, 2002; Zahra & McIntosh, 2007), (2) knowledge gain and increased skills (Blackman & Benson, 2010; Bruyere & Rappe, 2007; Campbell & Smith, 2006) and finally (3) enhanced interpersonal relations and social networks (McGehee, 2002; McGehee & Santos, 2005). For instance, five preferred outcomes of volunteer tourism experience are assessed by Bailey and Russel (2010): civic attitude, openness and a three-dimensional wisdom encompassing reasoning, emotions and contemplation. The authors opine that enhancement of these domains can enhance civic engagement and social capital.

Several authors postulate that transformative tourism experiences can positively impact lifestyles. Tourism is regarded as a healthful engaging pursuit that facilitates physical and psychological benefits. Travel offers an opportunity to visitors to pause and dissociate themselves from their daily mundane life and contemplate upon themselves and their overall lifestyle (Teoh et al., 2021). Furthermore, TE is influenced by numerous activators, progressions and outcomes. Few scholars have viewed TE from an existential perspective (Kirillova et al., 2017a, 2017b, 2017c; Pung et al., 2020).

For instance, Kirillova et al. (2017a) opine that epiphanic moments play a key role in transforming people. Walter (2016), however, uses a learning perspective to show that visitors "proactively self-examine and develop sustainable behaviors." Teoh et al. (2021) stress on a co-created approach that is focused on the tourists and the well-being of the host destination. The authors conceptualize the core of TE in a tourism place. The authors argue that transformative potential of tourism experiences is drawn from the landscape, social interactions and different ways in which an individual interplays with the society. Cavender et al. (2020) scrutinize an education abroad program based on Mezirow's transformational learning theory to understand the connotations and dimensions of transformative travel. Their focus is on the lived experiences of travelers in a program deliberately designed to aid in transformative learning. Some scholars postulate that if transformation learning programs are designed purposefully, they hold more meaning for the learners (Lange, 2004).

Sustained Transformation Catalysts

Furthermore, optimal flow, B-cognition moments or truth of the moment experiences often prompt transformation such as spontaneous and "meaningful events occurring towards the end of the trip, and characterized by intense emotions and a sense of precarious-ness, transcendence, and connection with nature or a divine figure" (Cavender et al., 2020, p. 4). Facilitators can be events and experiences which hold potential to induce a sense of place and generate captivating experiences which can captivate the attention of the guests as well as the hosts. The important task is to sustain these events; and for them to make an impressionable mark on the tourists, they need to self-sustain, be feasible and long-lasting. They need to reoccur and spark contemplation upon return home (Ponder & Holladay, 2013; Sheldon, 2020).

Focusing on design elements, Sheldon presents four scenarios that hold potential to transform consumers:

1 deep human connection in cultural contexts;
2 deep environment connection in natural contexts;
3 self-inquiry, self-reflection, learning and creativity and
4 engaged contribution to a destination by a tourist.

(2020, p. 10)

Sheldon also outlines five moments or triggers that relate to transformed consciousness:

1 admiration and wonder stimulated by natural magnificence and humble experience of one's insignificant position in the bigger order of things;

2 optimal experiences with deep emotions and disorienting predicaments evoking all senses;
3 flow;
4 mindfulness and
5 slow paced, uncomplicated and authentic experiences.

(Sheldon, 2020)

Pung et al. also place optimal experience and its distinct attributes as the basis of transformation and view "liminality, cultural shock, and physical and mental challenges as the triggering stimuli" (2020, pp. 8–10). The authors note that TEs can be prompted by interactions between hosts and tourists, disorienting experiences, a strong sense of place, extended stays and post-trip musings. However, potential to co-create transformation diminishes with brief stays, customary activities, traveling with familiar companions and lack of opportunity to be part of residents' daily life. Pung et al. (2020) posit that moments and experiences that promote eudaimonic (rather than hedonic) well-being hold potential to trigger sustained transformation in a tourist. In their study of a festival, Neuhofer et al. (2020) write that it is the positive emotions that result in sustained TEs and these can be triggered through design of creative narratives such as storytelling, generating themes, atmospherics and anesthetization.

Most scholars have posited that travel evokes the most favorable transformations in mindsets and values. It is also argued that mass tourism type of activities do not offer learning experiences with transformative potential. Instead, alternative forms of tourism with deeply immersive and learning opportunities (such as study abroad educational trips and activities) that are nature-based, culturally immersive, spiritual, volunteer, adventurous, and wellness-related (which include homestays) stimulate feelings of self-actualization and personal growth. Coghlan and Gooch (2011) refer to the learning phases of Mezirow to experiences that happen in a novel environment and stimulate deep emotions and desire to acquire new skills and develop social bonds such as in volunteer tourism. Outdoor and nature-tourism experiences generate a sense of awe and wonder of wilderness as tourists view first-hand damage to the wildlife and the natural landscapes in general. Activities designed as a 'nature shock,' 'adventure shock' and even 'culture shock' offer ideal opportunities for transformative learning, thereby cultivating contemplation toward significance of sustainable behavior. Pung et al. illustrate that "Mezirow's theory also helps explain tourists' experiences of culture shock at destinations: the cultural imbalance .... leads tourists to question subjective meanings and values, which encourages them to develop cross-cultural awareness: as travelers realise they have adapted their cultural constructs to different international realities, they transform .... by integrating elements of their experiences when returning to a 'blurred home' of new and old meanings" (2020, p. 4).

Pung et al. (2020) offer insights from an existential perspective, that is, they discuss how the self is involved in the transformation process by drawing on the manner in which Kirillova et al. (2017a, 2017b) explore TEs using an existential standpoint. It is argued that existential transformation leads to a change in a person's value system. A human being's existence is subject to change based on feelings of authenticity and anxiety. In other words, "anxiety is the fear of living a meaningless life and not expressing the best version of self (Kierkegaard, 1994). It pushes individuals to question their life's direction and values, this initiating a path towards transformation and achieving existential authenticity" (Kirillova et al., 2017b, 2017a, p. 501). According to these authors, a peak episode leads to a sudden revelation and triggers an existential transformation. It is described as a serendipitous moment that occurs toward the end of the trip (e.g. connecting with other people, reaching the top of a mountain and witnessing a striking local performance). Specifically, the peak experience, of a tourist, increases self-awareness and intensifies existential anxiety, leading to questioning of values and re-evaluation of priorities (Kirillova et al., 2017a). Existential transformation results in tourists achieving or enhancing their existential authenticity (Kirillova et al., 2017b). Existential authenticity, described as "potential existential state of Being which is to be activated by tourist activities" (Wang, 1999, p. 352), is perceived to be "difficult to obtain in daily life, due to ordinary routine and social pressures, which are suspended in the liminoid and self-explorative dimension of tourism settings" (Kirillova et al., 2017c, p. 17). Liminality refers to the ephemeral attributes of a place, which stimulate the tourists to let their guard down and connect with/express their intrinsic selves without any fear of being judged. For instance, it is the central part of a spiritual journey which offers tourists an opportunity to reassess and restructure the purpose of life and their existential self (Chhabra & Kim, 2024; Kirillova et al., 2017c; Timothy & Olsen, 2006; Willson et al., 2013).

Transformation Models

Pung et al. (2020) propose a tourist model of transformation where they show that theories such as cultural shock, disorienting dilemmas and peak moments/occurrences play a key role in starting the transformation process. This is followed by contemplation/meaning making reflection of the experience and results in enhancing existential authenticity or attainment of a new set of values such as responsible disposition toward the environment and mindfulness toward other cultures. The authors further offer a discursive view of how interaction with the tangible attributes of the visited destination triggers TEs. The performances and their expressions are regarded as important components of the TE. The key message the authors convey is that "a transformative experience is initiated by a stimulus, which tourists react to and interpret. The interpretation and integration of new knowledge and values may lead to attitude change and temporary habits; if these are enforced, long-term behavioral changes may be adopted" (Pung et al., 2020, p. 7). Drawing on Peirce's (Houser & Kloesel, 1992) theory of experience, the authors argue that intense stimuli trigger disruption of the self and existing values and stimulate the tourists to contemplate on their sense of self and its situatedness in the universe. That is, "Peirce's polar sense and Deleuze's fragmentation illustrate the state in which tourists are suddenly aware and upset by feeling an incongruence between their way of being and behaving at the destination and at home, in relation to the physical, psychological and sociological difference in their surroundings" (Pung et al., 2020, p. 8). Furthermore, sustained transformation is subject to the manner in which reminiscence happens upon return home. The depth of the reflection process would determine the extent of sustained transformation. As Pung et al. indicate, "if reinforced, temporary habits and attitude change" during the visit "can become long-term behavioral transformation at home, such as changing to new practices," such as environmental or conservation friendly behavior and/or personality/lifestyle changes (2020, p. 10).

Along similar lines, Teoh et al. (2021) posit that transformative learning experiences (TLEs) hold potential to shape lifestyles and therefore they can produce profound personal and social impacts on consumers. Using a hybrid organized narrative technique, the authors propose a holistic co-created transformational learning (TE) paradigm guided by three core domains: experience, experience facilitators and experience consumers. Facilitators (such as guests, hosts, tour guides, NGOs and researchers/educators) play a key role in orchestrating and shaping experiences of consumers. The in-situ cognitive interactions between consumers and facilitators hold potential to transform the facilitators as well as the consumers. The inter-connectedness between these dimensions generate four different types of outcomes which can be temporary or enduring. They can also be tangible or intangible. Tangible outcomes pertain to physical health appearance, whereas intangible refer to psychological transformations such as change in attitude or behavior. The former can be incidental, whereas the latter can induce enduring transformation.

The aforementioned studies offer important implications for tourism operators and/or destination marketing agencies as they share important information on how to shape tourism experiences as prospects to trigger an elevate state of consciousness and well-being. It is further suggested that the DMOs should: "curate their uniqueness, enhancing the awareness of the liminoid dimension of the destination setting, by promoting destination experiences of extraordinary natural sites and outstanding elements of local lifestyle (e.g. traditions, food, etc.) to increase the change of encounters with local residents, which could be effected through peer-to-peer guiding or home stays and to elicit tourist consciousness towards their embodied experiences through flow activities" (Pung et al., 2020. p. 10).

Cavender et al. (2020) present a YEP model based on three dimensions: 'yourself,' 'experience' and 'place.' 'Yourself' is directed toward inward focus and the traveler's 'Being in the World' before, during the journey and upon return home; their disposition to openly embrace new information and experience and get inspired. The experience theme is expressed using three sub-categories: (1) testing time (by going out of the comfort thresholds and experiencing fear and disoriented feelings due to unknown spaces which ultimately lead to learning, (2) exposure (to new information) and (3) making notes and contemplation (that is, penning down experiences assists holds potential to facilitate self-reflection upon return home). People and places theme in the YEP paradigm, however, are related to three sub-themes: explore, learn and care which galvanizes "a stronger understanding of the culture, the place and its people" and the travelers are dared to be watchful of their footprint on the place and the host community (Cavender et al., 2020, p. 10).

The Intrinsic Pathways

The eventual journey of all human beings follows an intrinsic path as indicated in numerous spiritual and philosophical philosophies (such as Taoism, Buddhism and the Hinduism-Vedanta). The final goal of this journey can be referred as "freedom, self-actualization, illumination, awakening, ultimate consciousness, or recognition of the divine" (Martin, 2019; Maslow, 1954; Sheldon, 2020; Tsu, 1972). The enlightened state of mind, whether it is temporary or permanent, "has been described as unity consciousness, transcendental consciousness, the flow experience, fundamental wellbeing, awakening, or non-dual awareness. Some traditions describe it as one moment of awakening where the egoic self drops away leaving the practitioner in a new and permanent state of enlightenment. Others see it as a lifelong path, requiring discipline and practice without any guarantee of achieving the final state" (Sheldon, 2020, p. 1). Tourism can serve as a conduit toward inner transformation for tourists and help to facilitate its realization.

Sheldon points out that "living life from a transformed or awakened state has profound significance for the individual, for society, and for tourism. This intrinsic journey of a person can be guided through numerous world traditions, ideologies and faiths" (2020, p. 2). The path of awakened consciousness inspires self-reflection, self-inquiry about the ultimate goal of life, abiding by elevated values and contributing to the well-being of the society. This sense of awakening is unique to each individual and helps in connecting with the true inner self (Coghlan & Weiler, 2018). According to Ross, the dynamic process of self-transformation can be captured in the following three phases:

a beginning with a disorienting dilemma and involves choice, healing and experience(s) of expanding consciousness toward the divine;
b initiating a permanent change in identity structures through cognitive, psychological, physiological, affective or spiritual experiences and
c rendering a sustained shift in the form of one's thinking, doing, believing, or sensing due to the novelty of the intersection between the experiencer, the experience, and the experiencer's location in time.

(2014, p. 54)

Popular aspects and merits of transformed consciousness include compassionate living, transcendence of self, altruist attitude by being mindful toward others and the environment, self-reflection, authenticity and simple living, and being mindful of the present moment (Sheldon, 2020). Furthermore, cultural immersions can stimulate deep human bonding. If retrospection stimulates deep and mindful admiration of other peoples' cultures, traditions, lifestyles and values/beliefs, transformational learning can be generated which can gratify tourists' quest for transcending the higher self (Filep et al., 2017; Kujawa, 2017; Sheldon, 2020). Interactions at the personal level can produce awakening moments and existentially authentic state of mind (Chhabra, 2021; Wang, 1999). If moments of exchange happen at a deeper and soulful level, they can produce long-term transformation. If the cross-cultural encounters are meaningful and if disorienting challenges generate an impact, there is more likelihood of transformation (Lyons, 2010; Tiberghien et al., 2020). Deep human bonds can generate affective bonds leading to transformed consciousness (Chhabra, 2022; Heimtun & Abelsen, 2012; Matteucci et al., 2022).

Transformative Spaces

Examples of settings and experiences that can enable interactive authentic moments and deep connectivity include storytelling (Soulard et al., 2019), creative episodes (Richards & Marques, 2012), or even situations where adversity or crisis exist; kindness and responsiveness of strangers can generate a bond (Filep et al., 2017). Research in this field is still scant. Glover and Filep (2017) stress on the significance of organizing meaningful meetings that are facilitated or fostered through spontaneous gestures of benevolence and compassion between guests and hosts. For example, the Appalachian Trail in the USA is famous for generating gestures of kindness and other types of inspiration for hikers. They happen unknowingly and through unfamiliar guests (Glover & Filep, 2017). Stories or narratives can strive to draw attention

FIGURE 2.1 Pristine natural environment

toward personal, cultural and historic diversity, or they can promote feelings of shared humanity and union between people (Sheldon, 2020). Flourishing destinations often draw wisdom from their communities and "stories and patterns of aliveness" hold potential to transform (Holliday, 2019, p. 1). Pristine natural environments, such as a river bank (see Figure 2.1), can facilitate deep enriching experiences. Landscapes and settings can be developed to encourage tourists and resident interactions through narration of deep personal stories that enhance the transformation potential of welcoming shared spaces in a destination. As an example, 'Holiday Participation Centers' are created for tourists in Flanders (Belgium) to share their narratives and contemplate on their trip from a meaningful and transformative vantage point (Holliday, 2019; Salazar et al., 2009). Local residents can also be invited to share their stories in their preferred way. It is important that design, events and interactions in shared or co-created spaces benefit local residents as well.

Sheldon envisages that creative spaces nourish deep human bonds. As an instance, designed experiences with the propose of engaging tourists in arts, music and/or performing arts serve as creative gates of entry in to the local community spaces and gratify the quest for transformation by engaging in creative and co-created activities (Richards & Marques, 2012). Furthermore, it is imperative to acknowledge the significant role of employees it triggering TEs. According to Sheldon, "employees at the service touchpoints of the tourist's experience can only contribute to the tourist's transformation if they have sophisticated communication and life skills, and have experienced some transformation of consciousness themselves. Then they can relate personally to the tourist and facilitate the transformative experience. The deep human connection and long-term relationships formed with employees can also contribute to the tourist's awakening" (2020, p. 6). Next section conceptualizes a transformative heritage tourism model.

Conceptualizing Transformative Heritage Tourism

Tasci and Godovykh (2021) write that sustainable tourism development embraces a more holistic approach today and has become multi-dimensional. It involves a broad spectrum of stakeholders albeit the wellness dimension has been mostly inspected from a commercial standpoint. Recently, a slow shift of focus has emerged toward the wellness of individuals and society, which has triggered a discourse on concepts such as happiness, well-being and transformation. Grounding their rationale in positive psychology, researchers call for tourism research and practice for transformative tourism experiences, which are not utilitarian or hedonic, but rather inner- and other-oriented and eudaimonic for potential transformation, well-being and happiness (Tasci & Godovykh, 2021).

Multiple pathways can realize the desire to enhance well-being. Theories such as flow experience theory and self-determination theory (SDT) have been employed extensively to obtain insights on depth of personal transformation during travel (Chhabra, 2021; Vada et al., 2020). Flow experience theory supports eudaimonic well-being as it refers to the extent to which personal skills can overcome activity-related dilemmas or challenges (Moneta & Csikszentmihalyi, 1996). A higher level of difficulty or disorientation can inspire contemplation toward the meaning and purpose of life (Tse et al., 2021). SDT refers to gratification of three needs: autonomy (sense of freedom), competence (mastering an activity or experience) and a sense of relatedness (social bonding) (Buzinde, 2020). Seligman's (2018) PERMA (positive emotions, engagement, relatedness, meaningfulness and accomplishment) model has also been used by several scholars to measure overall well-being (Chhabra, 2022; Dillette, 2016; Mendes et al., 2022). Huang et al. (2023) further this discourse examining the interrelatedness between needs and activities and well-being. They offer insights on need-fulfillment routes that tie with hedonic and eudaimonic well-being. Huang et al. (2023) propose a comprehensive paradigm synthesizing the processes underpinning distinct psychological corridors that connect activities, needs and hedonic/eudaimonic well-being.

McCabe and Johnson define well-being as "an integrative concept incorporating generic assessment of satisfaction with life overall, together with satisfaction in various domains of life, subjective assessments of emotional and social well-being, and assessments of psychological resources" (2013, p. 45). Well-being can be broadly delineated into two categories: hedonic (short-term gratification resulting in transitory joy or happiness and elevated mood) and eudaimonic ((self-realization and a higher state of consciousness).

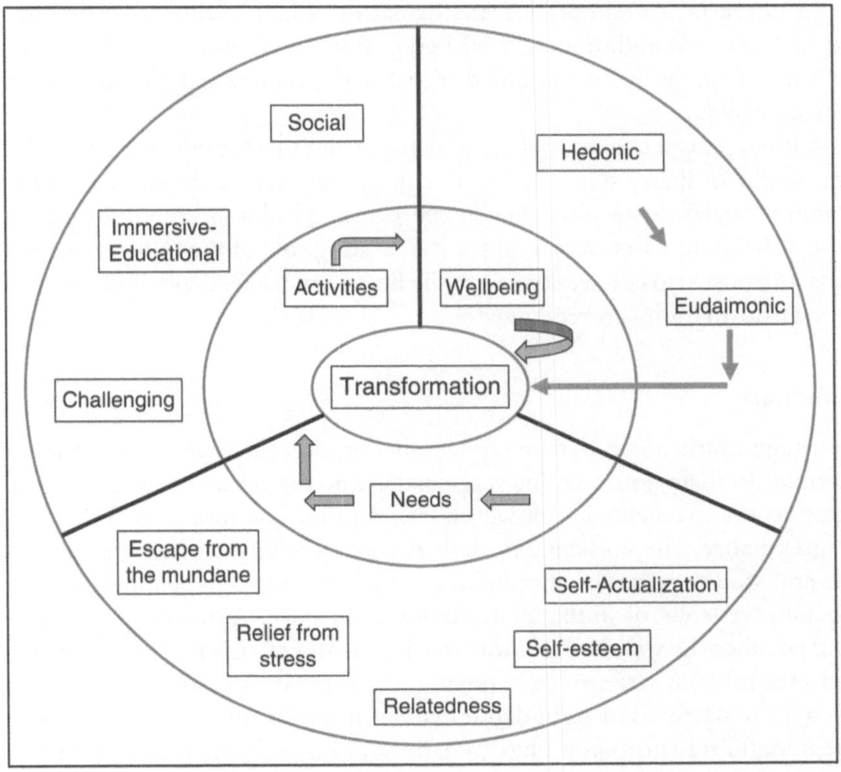

FIGURE 2.2 Conceptual path toward transformative heritage tourism

Specific activities and needs stimulate hedonic or eudaimonic well-being. According to Su et al. (2020), normally activities in the comfort zone stimulate hedonic well-being, whereas immersive and disorienting but meaningful experiences trigger eudaimonic well-being.

Figure 2.2 proposes a conceptual transformation model based on Huang et al.'s (2023) needs-activities-well-being sequential path of transformation. TEs can be realized through a set of niche forms of tourism such as adventure tourism, heritage tourism, volunteer tourism, backpacker tourism and wellness tourism as they offer opportunities to immerse/engage in the surroundings and new knowledge, foster deep bonds and experience optimal flow through disorienting or challenging activities and experiences. A broad spectrum of needs are captured in the model which align with several theories such as experience flow theory, SDT and Maslow's hierarchy of needs. A sequential path toward transformation is suggested in the model beginning with awareness of a need or set of needs (to seek deep experiences) and triggering the desire to travel and participate in activities (challenging, immersive, educational and social) that hold potential to stimulate transformation which ultimately contributes to hedonic and eudaimonic well-being. Pursuit of eudaimonic well-being can usher the path for sustained transformation toward a higher sense of consciousness.

Clearly, travel entices tourists to depart from their comfort zones and seek out-of-the-ordinary experiences. A continuum can be developed based on comfort-based versus disoriented experiences. The latter hold more potential to transform tourists. Based on the moderating role of the tourist's home setting and post-trip connections with the hosts and peers, depth and sustenance of transformation can be gauged.

Summary

Heritage tourism is a dynamic phenomenon. Recent trends associated with heritage tourism note ongoing surge in demand for authentic heritage tourism experiences. Authenticity, however, has multiple connotations. This chapter contextualizes the authenticity dialogue in the context of recent literature. Several studies postulate that heritage tourism, rooted in genuine or close to genuine versions of authenticity, holds tremendous potential to transform and enhance the well-being of tourists, hosts, and all other stakeholders in the heritage tourism system. Next, the notion of transformation and its semantic theory are narrated in an in depth manner in the chapter. It is explained that meaningful transformation has to happen at an individual level as well as at the systems level. A conceptual path toward sustained transformation of heritage tourism is recommended and the significance of eudaimonic well-being, to pave the path toward deep mindfulness and a higher sense of consciousness, is underlined. Chapter 3 by Xie focuses on heritage-based staycations as spaces of transformation. The author illustrates that staycations hold potential to benefit the local communities and offer a sustainable route to stimulate transformative heritage tourism.

References

Adams, V. (1996). *Tigers of the show and other virtual Sherpas: An ethnography of Himalayan encounters*. Princeton University Press.

Adie, B. A., & Falk, M. (2021). Residents' perception of cultural heritage in terms of job creation and overtourism in Europe. *Tourism Economics*, *27*(6), 1185–1201.

Agyeiwaah, E., Dayour, F., Otoo, F. E., & Goh, B. (2021). Understanding backpacker sustainable behavior using the tri-component attitude model. In H. Han (Ed.), *Sustainable consumer behaviour and the environment* (pp. 173–194). Routledge.

Alazaizeh, M. M., Ababneh, A., & Jamaliah, M. M. (2020). Preservation vs. use: Understanding tourism stakeholders' value perceptions toward Petra Archaeological Park. *Journal of Tourism and Cultural Change*, *18*, 252–266.

Apostolakis, A. (2003). The convergence process in heritage tourism. *Annals of Tourism Research*, *30*(4), 795–812.

Bailey, A., & Russel, K. (2010). Predictors of interpersonal growth in volunteer tourism: A latent curve approach. *Leisure Sciences*, *32*, 352–368.
Barbieri, N., & Marmol, C. (2022). Sustainability and commodification? The role of cultural assets in the development of new paths for tourism in Catalonia. In E. Borin, M. Cerquetti, M. Crispf, & J. Urbano (Eds.), *Cultural leadership in transition tourism: Developing innovative and sustainable models* (pp. 249–266). Springer International Publishing.
Belhassen, Y., Caton, K., & Stewart, W. P. (2008). The search for authenticity in the pilgrim experience. *Annals of Tourism Research*, *35*(3), 668–689.
Blackman, D. A., & Benson, A. M. (2010). The role of the psychological contract in managing research volunteer tourism. *Journal of Travel & Tourism Marketing*, *27*(3), 221–235.
Brooks, C., Waterton, E., Saul, H., & Renzaho, A. (2023). Exploring the relationships between heritage tourism, sustainable community development and host communities' health and wellbeing: A systematic review. *PLoS One*, *18*(3), e0282319.
Brown, E. J. (2013). *Transformative learning through development education NGOs: Comparative study of Britain and Spain* (Doctoral dissertation, University of Nottingham).
Brown, S., & Lehto, X. (2005). Travelling with a purpose: Understanding the motives and benefits of volunteer vacations. *Current Issues in Tourism*, *8*(6), 479–496.
Bruyere, B., & Rappe, S. (2007). Identifying the motivations of environmental volunteers. *Journal of Environmental Planning and Management*, *50*(4), 503–516.
Bueddefeld, J., & Duerden, M. D. (2022). The transformative tourism learning model. *Annals of Tourism Research*, *94*, 103405.
Buzinde, C. N. (2020). Theoretical linkages between well-being and tourism: The case of self-determination theory and spiritual tourism. *Annals of Tourism Research*, *83*, 102920.
Campbell, L. M., & Smith, C. (2006). What makes them pay? Values of volunteer tourists working for sea turtle conservation. *Environmental Management*, *38*, 84–98.
Carù, A., & Cova, B. (2003). Revisiting consumption experience: A more humble but complete view of the concept. *Marketing Theory*, *3*(2), 267–286.
Cavender, R., Swanson, J. R., & Wright, K. (2020). Transformative travel: Transformative learning through education abroad in a niche tourism destination. *Journal of Hospitality, Leisure, Sport & Tourism Education*, 27, 100245.
Chhabra, D. (2010). *Sustainable marketing of cultural and heritage tourism*. Routledge.
Chhabra, D. (2021). Transformative perspectives of tourism: Dialogical perceptiveness. *Journal of Travel & Tourism Marketing*, *38*(8), 759–768.
Chhabra, D. (2022). Transformational model of well-being for serious travellers. *International Journal of Spa and Wellness*, *5*(2), 129–146.
Chhabra, D., & Ashton, B. (2023). Marketing Authenticity Matters in Tourism. https://www.frontiersin.org/research-topics/53076/marketing-authenticity-matters-in-tourism
Chhabra, D., Healy, R., & Sills, E. (2003). Staged authenticity and heritage tourism. *Annals of Tourism Research*, *30*(3), 702–719.
Chhabra, D., & Kim, E. G. (2024). Sustaining inner transformation through spiritual tourism. *International Journal of Spa and Wellness*, 7(1), 43–64.
Chhabra, D., Sills, E., & Cubbage, F. W. (2003). The significance of festivals to rural economies: Estimating the economic impacts of Scottish Highland Games in North Carolina. *Journal of Travel Research*, *41*(4), 421–427.

Chung, J. Y., Kim, J. S., Lee, C. K., & Kim, M. J. (2018). Slow-food-seeking behaviour, authentic experience, and perceived slow value of a slow-life festival. *Current Issues in Tourism*, *21*(2), 123–127.

Coghlan, A., & Gooch, M. (2011). Applying a transformative learning framework to volunteer tourism. *Journal of Sustainable Tourism*, *19*(6), 713–728.

Coghlan, A., & Weiler, B. (2018). Examining transformative processes in volunteer tourism. *Current Issues in Tourism*, *21*(5), 567–582.

Collins-Kreiner, N. (2010). Researching pilgrimage: Continuity and transformations. *Annals of Tourism Research*, *37*(2), 440–456.

Cranton, P., & King, K. P. (2003). Transformative learning as a professional development goal. *New Directions for Adult & Continuing Education*, *2003*(98), 31–38.

Dans, E. P., & González, P. A. (2018). The Altamira controversy: Assessing the economic impact of a world heritage site for planning and tourism management. *Journal of Cultural Heritage*, *30*, 180–189.

DeSoucey, M., Elliott, M. A., & Schmutz, V. (2019). Rationalized authenticity and the transnational spread of intangible cultural heritage. *Poetics*, *75*, 101332.

Dillette, A. K. (2016). Wellness tourism: An application of positive psychological theory to overall quality of life (Doctoral dissertation, Auburn University).

Elabd, N., Mansour, Y., & Khodier, L. M. (2021). Utilizing innovative technologies to achieve resilience in heritage buildings preservation. *Developments in the Built Environment*, *8*, 100058.

Ensenat-Soberanis, F., & Blanco-Gregory, R. (2022). Crowding perception at the archaeological site of Tulum, Mexico: A key indicator for sustainable cultural tourism. *Land*, *11*, 1651.

Everett, S., & John Parakoottathil, D. (2018). Transformation, meaning-making and identity creation through folklore tourism: The case of the Robin Hood Festival. *Journal of Heritage Tourism*, *13*(1), 30–45.

Filep, S., Macnaughton, J., & Glover, T. (2017). Tourism and gratitude: Valuing acts of kindness. *Annals of Tourism Research*, *66*, 26–36.

Gino, F., Kouchaki, M., & Galinsky, A. D. (2015). RETRACTED: The moral virtue of authenticity: How inauthenticity produces feelings of immorality and impurity. *Psychological Science*, *26*(7), 983–996.

Glover, T. D., & Filep, S. (2017). Examining kindness of strangers in tourism: Trail magic on the Appalachian Trail. In S. Filep, J. Laing, & M. Csikszentmihalyi (Eds.), *Positive tourism* (pp. 135–147). Routledge.

Halewood, C., & Hannam, K. (2001). Viking heritage tourism: Authenticity and commodification. *Annals of Tourism Research*, *28*(3), 565–580.

Heimtun, B., & Abelsen, B. (2012). The tourist experience and bonding. *Current Issues in Tourism*, *15*(5), 425–439.

Heritage Tourism Outlook (2023). Heritage Tourism Outlook Report (2023-2033). Retrieved January 2024, from https://www.futuremarketinsights.com/reports/heritage-tourism-industry-outlook

Holliday, M. (2019). Travel to tomorrow: An emerging vision for the tourism industry, September 2019. https://michelleholliday.com/travel-to-tomorrow-an-emerging-vision-for-the-tourism-industry/, Accessed date: 9 December 2017.

Houser, N., & Kloesel, C. J. (Eds.). (1992). *The essential Peirce, volume 1: Selected philosophical writings (1867–1893)*. Indiana University Press.

Huang, X., Wang, P., & Wu, L. (2024). Well-being through transformation: An integrative framework of transformative tourism experiences and hedonic versus eudaimonic well-being. *Journal of Travel Research*, *63*(4), 974–994.

Huang, Y., Yi, D., & Clark, W. A. (2023). Subjective wellbeing in 21st century China: A multi-level multi-dimensional perspective on urban-rural disparities. *Applied Geography*, *159*, 103071.

Kierkegaard, S. (1994). *Fear and trembling and the book on Adler: Introduction by George Steiner*. Everyman's Library.

Kirillova, K., & Lehto, X. (2015). Destination aesthetics and aesthetic distance in tourism experience. *Journal of Travel & Tourism Marketing*, *32*(8), 1051–1068.

Kirillova, K., Lehto, X., & Cai, L. (2017a). What triggers transformative tourism experiences? *Tourism Recreation Research*, *42*(4), 498–511.

Kirillova, K., Lehto, X., & Cai, L. (2017b). Tourism and existential transformation: An empirical investigation. *Journal of Travel Research*, *56*(5), 638–650.

Kirillova, K., Lehto, X. Y., & Cai, L. (2017c). Existential authenticity and anxiety as outcomes: The tourist in the experience economy. *International Journal of Tourism Research*, *19*(1), 13–26.

Kujawa, J. (2017). Spiritual tourism as a quest. *Tourism Management Perspectives*, *24*, 193–200.

Lak, A., Gheitasi, M., & Timothy, D. J. (2020). Urban regeneration through heritage tourism: Cultural policies and strategic management. *Journal of Tourism and Cultural Change*, *18*(4), 386–403.

Lange, G. M. (2004). Wealth, natural capital, and sustainable development: Contrasting examples from Botswana and Namibia. *Environmental and Resource Economics*, *29*, 257–283.

Lean, G. L. (2009). Transformative travel: Inspiring sustainability. In *Wellness and tourism: Mind, body, spirit, place* (pp. 191–205). US: Cognizant Communication Corporation.

Leighton, D. (2007). Step back in time and live the legend: Experiential marketing and the heritage sector. *International Journal of Nonprofit and Voluntary Sector Marketing*, *12*, 117–125.

Lindberg, F., & Østergaard, P. (2015). Extraordinary consumer experiences: Why immersion and transformation cause trouble. *Journal of Consumer Behaviour*, *14*(4), 248–260.

Liu, Z., Zhang, M., & Osmani, M. (2023). Building information modelling (BIM) driven sustainable cultural heritage tourism. *Buildings*, *13*(8), 1925.

Lowenthal, D. (1985). *The past is a foreign country*. Cambridge University Press.

Lyons, T. (2010). Soft intervention and the transformation of militias into political parties. *Strengthening Peace in Post-Conflict War States*. (pp. 145–162).

Lyons, J. L. (2013). Autonomous cross-cultural hardship travel (ACHT) as a medium for growth, learning, and a deepened sense of self. In P. Burke (Ed.), *Cultural hybridity* (pp. 205–221). Routledge.

Mandić, A., & Kennell, J. (2021). Smart governance for heritage tourism destinations: Contextual factors and destination management organization perspectives. *Tourism Management Perspectives*, *39*, 100862.

Martin, J. (2019). *The finders*. Integration Press.

Maslow, A. (1954). *Motivation and personality*. Harper Publishing.

Matteucci, X., Volić, I., & Filep, S. (2022). Dimensions of friendship in shared travel experiences. *Leisure Sciences*, *44*(6), 697–714.

McCabe, S., & Johnson, S. (2013). The happiness factor in tourism: Subjective well-being and social tourism. *Annals of Tourism Research*, *41*, 42–65.

McGehee, N. (2002). Alternative tourism and social movements. *Annals of Tourism Research*, *29*, 124–143.

McGehee, N. G., & Santos, C. A. (2005). Social change, discourse and volunteer tourism. *Annals of Tourism Research*, *32*(3), 760–779.

McLean, F. (2002). *Marketing the Museum*. Routledge.

Mendes, J., Medeiros, T., Silva, O., Tomás, L., Silva, L., & Ferreira, J. A. (2022). PERMA model of well-being applied to Portuguese senior tourists: A confirmatory factor analysis. *Sustainability*, *14*(13), 7538.

Merriman, N. (1991). *Beyond the glass case: The past, the heritage and the public in Britain*. Leicester University Press.

Mezirow, J. (1991). *Transformative dimensions of adult learning*. Jossey-Bass.

Mkono, M. (2020). Eco-hypocrisy and inauthenticity: Criticisms and confessions of the eco-conscious tourist/traveller. *Annals of Tourism Research*, *84*, 102967.

Moneta, G. B., & Csikszentmihalyi, M. (1996). The effect of perceived challenges and skills on the quality of subjective experience. *Journal of Personality*, *64*(2), 275–310.

Moore, K., Buchmann, A., Månsson, M., & Fisher, D. (2021). Authenticity in tourism theory and experience. Practically indispensable and theoretically mischievous? *Annals of Tourism Research*, *89*, 103208.

Moscardo, G. (2001). Cultural and heritage tourism: The great debates. In B. Faulkner, G. Moscardo, & E. Laws (Eds.), *Tourism in the 21st century* (pp. 3–17). Continuum.

Neuhofer, B., Celuch, K., & To, T. L. (2020). Experience design and the dimensions of transformative festival experiences. *International Journal of Contemporary Hospitality Management*, *32*(9), 2881–2901.

O'Sullivan, D., & Jackson, M. J. (2002). Festival tourism: A contributor to sustainable local economic development? *Journal of Sustainable Tourism*, *10*(4), 325–342.

O'Sullivan, E. (1999). *Transformative learning: Education vision for the 21st century*. Zed Books.

O'Sullivan, E. (2002). The project and vision of transformative education: Integral transformative learning. In E. O'Sullivan, A. Morrell, & M. O'Connor (Eds.), *Expanding the boundaries of transformative learning: Essays on theory and praxis* (pp. 1–12). Palgrave Macmillan US.

Osello, A., Lucibello, G., & Morgagni, F. (2018). HBIM and virtual tools: A new chance to preserve architectural heritage. *Buildings*, *8*, 12.

Petronela, T. (2016). The importance of the intangible cultural heritage in the economy. *Procedia Economics and Finance*, *39*, 731–736.

Ponder, L. M., & Holladay, P. J. (2013). The transformative power of yoga tourism. *Transformational Tourism: Tourist Perspectives*, *1*(7), 98–108.

Poria, Y., Butler, R., & Airey, D. (2001). Clarifying heritage tourism. *Annals of Tourism Research*, *28*(4), 1047–1048.

Prentice, R. (1993). *Tourism and heritage attractions* (p. 253).

Pritchard, A., & Morgan, N. (2013). Hopeful tourism: A transformational perspective. In Y. Reisinger (Ed.), *Transformational tourism: Tourist perspectives* (pp. 3–14). CABI.

Pung, J., & Chiappa, G. (2020). An exploratory and qualitative study on the meaning of transformative tourism and its facilitators and inhibitors. *European Journal of Tourism Research*, *24*, 2404.

Pung, J. M., Gnoth, J., & Del Chiappa, G. (2020). Tourist transformation: Towards a conceptual model. *Annals of Tourism Research*, *81*, 102885.

Rastegar, R., Zarezadeh, Z., & Gretzel, U. (2021). World heritage and social justice: Insights from the inscription of Yazd, Iran. *Journal of Sustainable Tourism*, *29*, 520–539.

Rebec, K. M., Deanovic, B., & Oostwegel, L. (2022). Old buildings need new ideas: Holistic integration of conservation-restoration process data using heritage building information modelling. *Journal of Cultural Heritage*, *55*, 30–42.

Richards, G., & Marques, L. (2012). Exploring creative tourism. *Journal of Tourism Consumption and Practice*, *5*(2), 1–11.

Rickly, J. M. (2022). A review of authenticity research in tourism: Launching the annals of tourism research curated collection on authenticity. *Annals of Tourism Research*, *92*, 103349.

Rosenberg, M. J., & Hovland, C. I. (1960). Cognitive, affective, and behavioral components of attitude. In M. J. Rosenberg, C. I. Hovland, W. J. McGuire, R. P. Abelson, & J. W. Brehm (Eds.), *Attitude organization and change: An analysis of consistency among attitude components* (pp. 1–14). Yale University Press.

Ross, G. (2014). Meaning making in religious and spiritual travel. In S. Filep, & P. Pearce (Eds.), *Tourist experience and fulfillment: Insights from positive psychology*. Routledge.

Salazar, N. B., Van den Branden, E., Bryon, J., & Steylaerts, V. (2009). Cultural tourism storytelling in Flanders: the story behind the stories.

Sands, R. (2023).: History Travel Offers Authentic Heritage Tourism Experiences. Retrieved January 2024, from https://www.forbes.com/sites/rogersands/2023/02/20/history-travel-offers-authentic-heritage-tourism-experiences/?sh=526a813b20e6

Scarpi, D., & Raggiotto, F. (2023). A construal level view of contemporary heritage tourism. *Tourism Management*, *94*, 104648.

Seligman, M. (2018). PERMA and the building blocks of well-being. *The Journal of Positive Psychology*, *13*(4), 333–335.

Sevieri, G., & Galasso, C. (2021). Typhoon risk and climate-change impact assessment for cultural heritage asset roofs. *Structural Safety*, *91*, 102065.

Sheldon, P. J. (2020). Designing tourism experiences for inner transformation. *Annals of Tourism Research*, *83*, 102935.

Snowball, J. D. (2013). The economic, social and cultural impact of cultural heritage: Methods and examples. In I. Rizzo & A. Mignosa (Eds.), *Handbook on the economics of cultural heritage* (pp. 438–455). Edward Elgar Publishing.

Soulard, J., McGehee, N. G., & Stern, M. (2019). Transformative tourism organizations and glocalization. *Annals of Tourism Research*, 76, 91–104.

Soulard, J., McGehee, N. G., Stern, M. J., & Lamoureux, K. M. (2021). Transformative tourism: Tourists' drawings, symbols, and narratives of change. *Annals of Tourism Research*, *87*, 103141.

Su, L., Tang, B., & Nawijn, J. (2020). Eudaimonic and hedonic well-being pattern changes: Intensity and activity. *Annals of Tourism Research*, *84*, 103008.

Tasci, A. D., & Godovykh, M. (2021). An empirical modeling of transformation process through trip experiences. *Tourism Management*, *86*, 104332.

Taylor, J. (2001). Authenticity and sincerity in tourism. *Annals of Tourism Research*, *28*(1), 7–26.

Taylor, E. W. (2007). An update of transformative learning theory: A critical review of the empirical research (1999–2005). *International Journal of Lifelong Education*, *26*(2), 173–191.

Teoh, M. W., Wang, Y., & Kwek, A. (2021). Conceptualising co-created transformative tourism experiences: A systematic narrative review. *Journal of Hospitality and Tourism Management*, *47*, 176–189.

Tiberghien, G., Bremner, H., & Milne, S. (2020). Authenticity and disorientation in the tourism experience. *Journal of Outdoor Recreation and Tourism*, *30*, 100283.

Timothy, D. J. (2018). Making sense of heritage tourism: Research trends in a maturing field of study. *Tourism Management Perspectives*, *25*, 177–180.

Timothy, D. J. (2020). Heritage consumption, new tourism and the experience economy. In M. Gravari-Barbas (Ed.), *A research agenda for heritage tourism* (pp. 203–218). Edward Elgar Publishing.

Timothy, D. J. (2021). *Cultural heritage and tourism: An introduction* (2nd ed.). Channel View Publications.

Timothy, D. J., & Boyd, S. W. (2003). *Heritage tourism*. Prentice Hall.

Timothy, D. J., & Olsen, D. H. (Eds.). (2006). *Tourism, religion and spiritual journeys* (Vol. 4). Routledge.

Trope, Y., & Liberman, N. (2010). Construal-level theory of psychological distance. *Psychological Review*, *117*(2), 440.

Tse, D. C., Nakamura, J., & Csikszentmihalyi, M. (2021). Living well by "flowing" well: The indirect effect of autotelic personality on well-being through flow experience. *The Journal of Positive Psychology*, *16*(3), 310–321.

Tsu, L. (1972). *Tao Te Ching* (G. Feng & J. English, Trans.). Vintage.

Ulusoy, E. (2016). Experiential responsible consumption. *Journal of Business Research*, *69*(1), 284–297.

Vada, S., Prentice, C., Scott, N., & Hsiao, A. (2020). Positive psychology and tourist well-being: A systematic literature review. *Tourism Management Perspectives*, *33*, 100631.

Walter, P. G. (2016). Catalysts for transformative learning in community-based ecotourism. *Current Issues in Tourism*, *19*(13), 1356–1371.

Wang, N. (1999). Rethinking authenticity in tourism experience. *Annals of Tourism Research*, *26*(2), 349–370.

Wearing, S. (2002). Re-centring the self in volunteer tourism. In G. Dann (Ed.), *The tourist as a metaphor of the social world* (pp. 237–262). Cabi Publications.

Wearing, S., & Neil, J. (2001). Expanding sustainable tourism's conceptualization: Ecotourism, volunteerism and serious leisure. In S. McCool & R. Moisey (Eds.), *Tourism, recreation and sustainability: Linking culture and the environment* (pp. 233–254). Cabi Publishing.

Willson, G. B., McIntosh, A. J., & Zahra, A. L. (2013). Tourism and spirituality: A phenomenological analysis. *Annals of Tourism Research*, *42*, 150–168.

Xiao, W., Mills, J., Guidi, G., Rodríguez-Gonzálvez, P., Gonizzi Barsanti, S., & González-Aguilera, D. (2018). Geoinformatics for the conservation and promotion of cultural heritage in support of the UN sustainable development goals. *ISPRS Journal of Photogrammetry And Remote Sensing*, *142*, 389–406.

Yang, J. Y., Paek, S., Kim, T., & Lee, T. H. (2015). Health tourism: Needs for healing experience and intentions for transformation in wellness resorts in Korea. *International Journal of Contemporary Hospitality Management*, *27*(8), 1881–1904.

Yi, X., Fu, X., Lin, B., & Sun, J. (2024). Authenticity, identity, self-improvement, and responsibility at heritage sites: The local residents' perspective. *Tourism Management*, *102*, 104875.

Yi, X., Fu, X., Lin, V. S., & Xiao, H. (2022). Integrating authenticity, well-being, and memorability in heritage tourism: A two-site investigation. *Journal of Travel Research*, *61*(2), 378–393.

Yorks, L., & Kasl, E. (2006). I know more than I can say: A taxonomy for using expressive ways of knowing to foster transformative learning. *Journal of Transformative Education*, *4*(1), 43–64.

Zahra, A., & Mcintosh, A. J. (2007). Volunteer tourism: Evidence of cathartic tourist experiences. *Tourism Recreation Research*, *32*(1), 115–119.

Zhang, Z., & Zou, Y. (2022). Research hotspots and trends in heritage building information modeling: A review based on CiteSpace analysis. *Humanities and Social Sciences Communications*, *9*, 394.

Zhenrao, C., Chaoyang, F., Qian, Z., & Fulong, C. (2021). Harmonious Coexistence of Cultural Heritage Protection and Tourism: The Mount Lushan Comprehensive Tourism Platform. *Heritage Science*, *9*(1): 1–16.

3

HERITAGE TOURISM AS TRANSFORMATION IN STAYCATION ERA

Philip Xie

Introduction

Heritage planning has been used interchangeably in tourism studies referring to conservation or regeneration (Janssen et al., 2014; Kalman, 2014). It essentially differs from the preservation discourse in terms of the conceptualization of heritage for tourism (Chhabra, 2021; Timothy & Boyd, 2003). The scope of heritage planning expands to a much larger picture of the past, going beyond recognized heritage sites and material remains and becoming sensitive to diverse interests in interpreting and representing cultural heritage (Van Assche & Duineveld, 2013, p. 2). In a similar vein, heritage tourism moves from "historical accuracy" and "intrinsic authenticity", to "the contemporary extrinsic narrative attached to it" (Patiwael et al., 2019, p. 336). It appears that heritage tourism in contemporary society requires planners and professionals to change from experts to mediators and facilitators and create a supportive environment to involve the public in heritage management.

Concomitantly, the global pandemic of COVID-19 starting 2020 upends travels. The notion of staycation has increasingly become an alternative when visitors are physically restrained to long-haul travels (Higgins-Desbiolles, 2020). The "staycation" is a neologism coined from "stay-at-home" and "vacation", loosely defined as "a vacation in which the vacationer stays at home, or near home, while creating the environment of a traditional vacation" (Wixon, 2009, p. 2). Similar terms, such as nearcation, holistay, and minibreak, gradually gained attention during the global financial crisis in 2008 when many residents chose to travel close to home (James et al., 2017). Staycation gradually morphs into a complex cultural phenomenon (Sharma, 2009), including a changing notion of modernity for "slow travel" social movement (Molz, 2009) as well as a prominent case of "proximity tourism"

DOI: 10.4324/9781003346197-3

(Jeuring & Haartsen, 2017). More importantly, staycation turns into a pathway for a new form of heritage tourism (Bafadhal, 2021; Besson, 2017).

Previous studies (Alejziak, 2013; Alexander et al., 2011) reveal that a staycation encourages people to travel locally at a steady cost. It is increasingly viewed as a necessary form of mobility largely defined by government's crisis regimes of (im)mobility, enforced through decrees and explained via ever-changing guidelines (Salazar, 2021). For example, the Ontario Government in Canada has initiated "staycation tax credit" to boost local business by offering people who book overnight stays in Ontario for anytime in 2022 a return of 20% on accommodation expenses of up to $1,000 per person or $2,000 per family (Xing, 2022). Similarly, staycation offers a vicarious travel experience that perceived heritage is deemed as an essential construct. Therefore, this study aims to identify what and how residents experienced heritage tourism program offered by local government in staycation era; specifically, it reveals that heritage plays a unique role for locals, while staycation serves a different purpose for its planning.

The city of Macau has been chosen for this study, which has been a colony under Portuguese rule for more than 400 years until it was handed over to the People's Republic of China in 1999. Macau was far less appealing as a tourist destination, but things quickly changed when the government liberalized its casino industry in 2002 and aggressively positioned Macau as the world's casino capital (Lou & Xie, 2018). However, starting in January 2020, due to COVID-19 outbreak, strict travel restrictions, especially the closure of the borders to Mainland China and Hong Kong, have shuttered the hospitality and gaming industry. With its heavy dependence on Mainland Chinese tourists, the number of arrivals declined drastically at the rate of 90% in 2020. As more than 80% of local population is working in gaming, tourism, and hospitality industry, the pandemic brings major economic artery to a standstill and exposes challenges for economy and livelihood. Hundreds of hotels and casinos are on the brink of shutting down their operations.

The staycation rejuvenation package entitled "Macao Ready Go" was initiated and subsidized by Macau government in June 2020 to support local tourism initiatives and reinforce the local economies. They encompass heritage-based tours designated to learn about the history of Macau as expressed in archaeological and industrial sites, museums, Macanese cuisine, and colonial architecture. In addition, resort and nature-based routes include tours of casino and sightseeing to explore the rich historical, cultural, and natural heritage. The chapter begins with an introduction to the notion of staycation. The focus is on heritage-based tours, comparing with resort and nature-based ones to better understand how to transform conventional heritage planning amid the global pandemic. Methodology and findings are reported, and the conclusions are discussed in the end to help tourism marketers to formulate future strategies in promoting their local services.

Heritage Tourism and Staycation Programs

The salient feature of postmodernism is that travel and tourism are assumed to be largely equivocal and subjective: heritage is not a concrete entity acted upon by forces from outside, but rather as sets of symbols, or as webs of significance and meaning (Xie, 2015). Waterton and Watson (2013) identify the issues of heritage and propose three frames to address the transformation of heritage tourism, e.g., theories *in*, *of*, and *for* heritage. Theories *in* heritage largely concentrate on "best practice". The conceptualization of heritage in architecture, fine arts, and related discipline applies this frame to an instrumental use of heritage in various studies (Mason, 2004). Theories *of* heritage shift attention from substance operationalization to the interpretation of the sociocultural process. The frame focuses on the representation of meaning as a discursive approach to decode heritage concept. Lastly, theories *for* heritage aim to respond daily experience and practice by human beings. They fit what French philosopher Michel de Certeau (2011) termed *L'invention du quotidien* (the practice of everyday life) so that facets of social forms such as traditions, heritage, and art make up a culture. Since culture is being constantly appropriated and consumed by people in everyday situations. Certeau's approach to "quotidian" as the system and discipline of the status quo is discernible in modern society.

The COVID outbreak in 2020 reveals a new form of heritage tourism, which rests upon the capability of "the infinite quotidian reproducibility" (Anderson, 1991, p. 183) and "ways of operating" (Certeau, 2011: xviii). Staycation erases the chaos of everyday life and replaces it with recognizable markers. The locals allude to the creations of heritage sites in size, scope, architectural detail, and meaning. However, the representations typically are not exclusively from a particular community's history because the staycation planners formulate a version of consumable heritage by selecting various images and symbols, by rejecting some and by creating new components. Heritage-based tours in the context of staycation necessarily falsify place and time by inventing an ideal setting where "space [is] practiced place" (Certeau, 2011, p. 117). The concept of dissonant heritage in staycation era is largely replaced by an array of tourism "*practiques*": the aestheticization of space, the construction and production of nostalgia, and the idealization of colonial history. The value attached to staycation programs motivates locals to seek out distinct cultural elements, such as rituals, myths, customs, dress, beliefs, and food.

Crisis engenders different understandings from both lexical and conceptual aspects (Koselleck, 2006). Staycation occurred during COVID-19 outbreak reflects a time of intense difficulty, as well as an urgent need to supply psychological restorative resources. Previous studies of staycation illuminate a pattern when residents prefer using discretionary time to rediscover their local environment, such as going to a variety of establishments with a short drive

(Ashbullby et al., 2013; MacKerron & Mourato, 2013; Pyke et al., 2016). Likewise, Larsen (2008) calls staycation as a form of "de-exoticization of tourist travel". He argues for the assumption that tourism is primarily about experiencing the extraordinary or exotic elsewhere in contrast to the everydayness of home geographies. In this case, staycation is more about the reproduction of familiar social relations than an escape from social obligations. Molz (2009) suggests that the lure of staycation tends to be associated with positive values such as freedom and progress in a familiar environment. Gonçalves-Sá (2020) exemplifies that nature-based route, such as hiking on local parks or staying a nearby hotel, is ideal to spend leisure times. However, heritage-based activities, such as visiting local museums, vernacular architecture, or historic districts, are frequented by staycationers to relive the best parts of their hometown (Andriotis, 2018; Chen & Chen, 2010). Although nature and heritage-based attractions are the mainstay of staycation programs (Mehmetoglu, 2007), the emerging health-oriented staycation is gaining popularity whose primary purpose focuses on mental health and psychological well-being (James et al., 2017). Therefore, staycation has increasingly become a byproduct of travel restrictions when people realign and change their destination choices.

The staycation programs entitled "Macao Ready Go! Local Tours", coordinated by the Macao Government Tourism Office (MGTO), were chosen as research subjects. The project was rolled out to offer 25 themed programs targeting residents between June and September 2020, which aim to form strategic alliances with tourism and hospitality sectors while keeping the industry afloat and restore confidence in Macau as a tourist destination. The variety of itineraries offer participants a chance to visit beautiful nature spots, cultural heritage sites, museums, and parks, as well as enjoy the city's integrated resorts. In particular, the programs offered in partnership with the city's historic, cultural, outdoor recreation organizations. Historians, naturalists, and tour guides lead participations on explorations through historic districts, behind-the-scenes cultural site tours, nature walks, trail hikes, and tours of many of the industrial sites that were once thriving local industries. These programs strictly followed government regulations, such as temperature checks and hand-sanitizer dispensers.

Provisional figures show that close to 140,000 staycationers participated in these designated programs (MGTO, 2020). The programs broadly encompass both "heritage-based tours" and "leisure tours". Approximately 60% of staycationers opted for "leisure tours", whereas 40% chose heritage-based tours. Among these programs, the "teamLab SuperNature" tour attracted over 37,000 participants, the highest ranking of "leisure tours", while over 15,000 signed up for the "maritime tour" for the colonial history of Macau, the most popular one in "heritage-based tours". Other popular activities include "experience Macau by land, sea and sir" (about 19,000 participants), "love the Country, love Macao" (about 9,800 participants), and "a field trip to the Grand Coloane resort" (about 8,500 participants) (ibid., 2020).

Since the staycation programs were designed for Macau residents only, they were required to sign up through travel agencies. A participant can choose one of either the "heritage-based tours" and/or "leisure tours". The government pays a subsidy of 280 MOP (US$35) per trip, to a maximum of 560 MOP (US$70) (Moura, 2020). The "Macao Ready Go" was tailored to provide residents with economical travel plans and opportunities to get familiarized with the latest development of tourism products.

Methodology

The survey questionnaire was developed based on extensive literature review. A total of 41 questions were designed to construct the conceptual model. The tours are subcategorized in three major foci, e.g., nature, heritage, and resort-based staycations. A pilot study was performed to ensure that the respondents understood the content of the questions and the description of the staycation experience. The pilot study also covers an examination on wording, composition, and sequence of questions. The survey utilizes a five-point Likert-type scale, ranging from 1 (totally disagree) to 5 (totally agree).

Macau residents can sign up at the accredited travel agencies in the city to book a government-subsidized staycation. A total of 317 valid questionnaires were filled out via both online and offline travel agency platforms. Among them, 108 responses are from heritage-based, 109 from nature-based tours, and 100 from resort-based ones. The convenience sampling was adopted to ensure the reliability and validity of the survey. For the on-site survey, it was conducted through the tour guides who led staycation groups and distributed the self-administered questionnaires when the program was completed. The survey process was undertaken from September to October 2020. On average, approximately 52 staycationers per week were surveyed during that period and 257 valid responses were collected. Meanwhile, the travel agencies offered online survey for those participated in staycation, including heritage-, nature-, and resort-based programs. The participants were asked to specify the type of programs first and their experiences in writing.

This study undertook SPSS 20.0 to analyze collected data. Factor analyses with both exploratory and confirmatory aspects were employed to identify the construction of the staycation experience. In addition, ANOVA tests were operated to better understand the difference among three major staycation programs.

Findings

Demographic Profiles of Respondents

The staycationers' sociodemographic information is presented in Table 3.1. Female respondents showed the highest interest (60%) in resort-based

TABLE 3.1 Demographic Characteristics of Respondents (N = 317, Unit: %)

Factors	*Nature-based Group*	*Heritage-based Group*	*Resort-based Group*	*Factors*	*Nature-based Group*	*Heritage-based Group*	*Resort-based Group*
Gender				*Marital status*			
Female	50	45	60	Married	40	62	72
Male	50	55	40	Single	60	38	28
Age				**Education background**			
18–25	17	16	29	Junior school	24	15	19
26–35	45	30	40	High school	42	28	37
36–45	20	35	13	College graduation	20	26	24
46–55	14	14	10	Bachelor's degree	13	25	19
Above 56	4	10	8	Master's degree and higher	1	6	1
Occupation				**Monthly income (MOP)**			
Civil officer	10	10	0	Less than 10,000	5	3	2
Hospitality employee	27	24	19	10,001–15,000	12	8	9
Teacher	7	13	2	15,001–20,000	25	29	28
Doctor	12	5	2	20,001–25,000	27	27	30
Lawyer	5	5	2	25,001–30,000	20	22	26
Self-employed	18	15	25	Above 30,000	11	11	5
Retired	6	10	19				
Student	9	12	24				
Others	6	6	7				

staycation, while males prefer heritage-based tours (55%). The majority of staycationers are in the age group of 26–35, indicating that these programs were popular among the younger residents. With respect to the occupation, hospitality-sector employees were the majority as gaming and casino industry takes the lead in Macau. It also shows that they were willing to participate in staycation programs due to furloughs and no-paid leave. Ironically, this group was not interested in taking the resort-based tours which happened to be their working places.

In terms of marital status, the married expressed a strong interest in heritage- and resort-based tours, mainly because staycation promotes a sense of gathering where heritage and/or resort programs are ideal places for families and married couples. More than 50% of respondents earned high school degrees when reporting educational attainment. This is largely attributed to the local government policy to support residents to work in casino once they graduate from high school. In terms of their income, the participants with relatively lower income which is below 15,000 MOP (US$1,875) are fewer than other respondents to join in this staycation. Current pandemic has impacted tourism and hospitality industry in Macau which is difficult for the lower-income group to participate in staycation despite the subsidies offered by the local government.

Factor Analysis of Staycation Experience

Exploratory factor analysis was initially undertaken to identify the attributes of staycationers' experience. The Kaiser-Mayer-Olkin (KMO) measure of sampling adequacy value is 0.851, while the Bartlett's Test of Sphericity shows 1107.5 ($p < .001$) indicating that the sample is appropriate for an exploratory factor analysis. Specifically, two of the items in the scale of staycation experience were deleted as the rotated loading demonstrates lower than 0.5. As a result of population attributable fraction (PAF) along with the varimax rotation, three factors within selected nine items were statistically significant, including peace of mind, pleasant surprise, and hedonic happiness that explain 78.5% of the total variance. As shown in Table 3.2, all the item loadings exceed the recommended value of 0.5, the construct CR values are greater than 0.7, and the AVE values exceed the threshold value of 0.5 (Byrne, 2013). Therefore, convergent validity of the staycationers' travel experience is well-established in these samples.

Discriminant validity is assessed by using a Fornell-Larcker (F-L) criterion analysis (1981) and Heterotrait-Monotrait ratio (HTMT). The F-L criterion suggests a higher coefficient of each construct (Hair et al., 2010). In

TABLE 3.2 The Construction of Staycation Experience

Constructs of Staycation Experience (SE)	*Means*	*Loading*	*C.R.*	*AVE*	*Total Variance Explained*
SE1-Peace of Mind			0.89	0.73	34.250
I had a feeling of being safe and protected	4.86	0.79			
I felt inner calm when participating in staycation	4.57	0.73			
I felt relaxed and a change of pace	4.29	0.73			
SE2-Pleasant Surprises			0.88	0.79	35.048
I was happy about an unexpected heritage event	4.36	0.85			
I was enchanted by cultures I revisited	3.92	0.68			
I was delighted by service and programs	4.06	0.79			
SE3-Hedonic Happiness			0.87	0.71	27.585
Happiness is quickly achieved through enjoyment	4.58	0.81			
My imagination jumped over the moon	3.27	0.89			
I'm overjoyed with space and comfort	4.86	0.77			

$\chi2/df = 2.957$; CFI = 0.922; GFI = 0.945; AGFI = 0.938; RMSEA = 0.062; RMR = 0.081

addition, there is not any HTMT ratio higher than 0.9, suggesting that the variables used in this study are valid (Henseler et al., 2015). Furthermore, there is no multicollinearity issue between the construct items. Table 3.2 also shows that the result of RMSEA is 0.062 and that all the constructions of staycation experience meet the requirements for goodness-of-fit (Henseler & Sarstedt, 2013).

The mean values of each category demonstrate positive perception and reflective experience of staycation in lieu of long-distance journey. For example, the items for "peace of mind", such as feeling safe and protected, show the highest mean value (4.86), while the "inner calm" achieved the second highest at 4.57. It is evident that staycation becomes a viable path for releasing stress and achieves subjective well-being for those suffering from global pandemics. This is particularly important for heritage tourism where learning for different culture is replaced by the factors of peace of mind and inner calm to explore its own culture. The peace of mind is also derived from avoiding some of the most frustrating hassles associated with traveling abroad. However, pleasant surprises reflect a state of being to relive the memories of familiar environment in an unexpected way. The items such as "happy about an unexpected heritage event", "enchanted by scenes and cultures I revisited", and "I was delighted by service and programs" recorded the higher mean value. However, hedonic happiness such as "my imagination jumped over the moon" received the lowest mean value (3.27). This explains that the notion of hedonic happiness tends to focus on the rising awareness of the extra space and comfort offered by different programs (4.86). The "wow" factors play a key role in staycation planning when residents actively seek to discover something unique and different, including heritage-based programs.

The factor analysis shows that restorativeness becomes an important attribute to recondition stress environment and solve fatigue problems such as attention dispersion, irritation, and agitation. Heritage tourism is replaced by a sense of psychological restoration ameliorating negative effects on the restricted environment (Lin et al., 2021), such as lockdowns and quarantines during COVID outbreak. Kaplan (1995) uses the attention restoration theory to further explain the restorativeness that can be measured in terms of four characteristics, e.g., being away, fascination, extent, and compatibility. Among these, being away is equivalent to a sense of physical or mental distance that demands the consumption of directed attention, including both physical and notional changes away from daily routines (Lee & Hyun, 2003). Fascination is the central component of restorativeness referring to the core of recreational activities such as visiting heritage sites or hiking in a natural environment. The notion of extent indicates a mental extension created through the constituents of the natural/built environment (Ko, 2013), which is considered to experience the environmental harmony between leisure space and internal/external elements of the landscape. Lastly, compatibility is predicated within the environment linking the level of correspondence to an individual's demand.

Comparative Study of Staycation Programs

The difference among three major types of staycations, e.g., heritage-, nature-, and resort-based programs, is compared by using a series of ANOVA. As shown in Table 3.3, three programs differ statistically from peace of mind, pleasant surprise, and hedonic happiness ($p < .001$). Among these, the attribute of "hedonic happiness" ($F = 8.51$) exhibits the divergence of three programs, along with "peace of mind" ($F = 6.27$). It appears that the interpretation of hedonic happiness varies significantly, while joys occurred in the context of heritage-based than nature- and resort-based ones. Staycations allow residents to see the city in a different light, steering away from what they usually do and where they usually go. The changing environment in staycation plays a key role in bringing a sense of happiness and peace of mind to locals. It appears that staycation in the resorts shows the highest level of self-worth, largely due to the uniqueness of Macau where most participants are also the employees of those resorts; the role change from serving guests to be served clearly enhances their self-identity.

In staycation era, the concept of heritage moves from monumental and aesthetic preferences to a complexity of physical, discursive, and affective dimensions. The findings echo the previous study of staycation consumption (Moon & Chan, 2022) that locals face (disillusion of traveling) and from their own fantasy (illusion of traveling). Staycation experience dynamically comprises their physical and cognitive activities, social interactions, and emotional responses, whereby they gain a feeling of extraordinariness combined with the ordinariness of familiar surroundings. In addition, staycation experiences are underpinned by a full mediating effect between place attachment and experience quality which is performed by sense of presence, together with consolidation of experience quality and psychological detachment from home base (Yan et al., 2022).

TABLE 3.3 ANOVA Analysis Result

Staycation Experience	*Types of Staycations*			*F*	*Sig. Nature*
	Nature	*Heritage*	*Resort*		
SE1-Peace of mind	4.12	4.26	4.39	6.27	<.001
SE2-Pleasant Surprise	3.79	4.08	3.92	4.09	<.001
SE3-Hedonic Happiness	3.85	4.04	3.86	8.51	<.001

Research Implications

The global pandemic that emerged at the beginning of 2020 has brought the tourism industry to a screeching halt and destinations reliant on tourists have been the hardest hit. The COVID-19 also causes a riptide of fear and confusion, dragging the whole travel industry down. However, staycation has been developed as a means of economics heavily promoted by government. This study used Macau as the case study to discuss staycation-based heritage planning during COVID outbreak. It seems that heritage planning in staycation era disconnects from the progress made by the official heritage framework. Heritage tourism serves a different purpose in staycation era when peace of mind, pleasant surprises, and hedonic happiness play a key role in transforming heritage to public consumption.

The study is exploratory in nature and specific in Macau where staycationers work in tourism and hospitality sectors and their livelihood is closely tied to travel industry. This is totally different from previous studies where many participants had nothing to do with tourism and hospitality fields (James et al., 2017). Second, the sociodemographic suggests a distinct pattern as a younger female group, married with relative high income, is the mainstay of the staycation. The group is heavily involved with hospitality sectors in Macau and has a prior knowledge in choosing the type of recreational activities; therefore, their feedback clarifies expectations and improves the future development of staycation. Lastly, heritage tourism in the context of staycation demonstrates a process of placemaking, which intentionally leverages the power of the arts, culture, and creativity to serve a community's interest while driving a broader agenda for improving local economy.

Staycation tends to serve as a compensatory means for those who can't participate in long-haul travels and an alternative for choosing heritage and leisure locally (Alejziak, 2013; Higgins-Desbiolles, 2020). Three specific attributes, e.g., peace of mind, pleasant surprise, and hedonic happiness, turn staycation to a rewarding experience and a state of restorativeness Spending quality time with family and friends away from the stresses of work is the respite many staycationers need in a difficult time. The programs supported by the local government stimulate peace of mind and hedonic happiness when local attractions satisfy residents' travel motivations. To some extent, staycation becomes a vicarious authentic experience for heritage-, nature-, and resort-based tourism when travel restrictions remain in place.

The findings resonate to previous studies of staycation where it becomes a vital tool for heritage tourism development (Cho et al., 2016; Laumann et al., 2001; Marselle et al., 2016). The function of staycation reveals self-actualization of discretionary time, eventually leading to community attachment, relaxation, and a sense of authenticity. With respect to the structural link of the staycation experience and its key attributes, the findings reveal that heritage is camouflaged in the leisure context where the pursuit of restorativeness is deemed important. The primary purpose of staycation becomes an essential mobility supported by government to bolster the tourism and hospitality sector. It is built on a wide range of government initiatives designed to help these sectors through the COVID pandemic and makes travels more affordable in the near future.

Finally, heritage-, nature-, and resort-based routes show a different response through the ANOVA analysis. Staycationers in Macau clamor for experiences over amenities, such as maritime activities, spa, and family events situated in a familiar setting. The heritage-based tours fit the tourist city like Macau which owns many colonial buildings and history of architecture. Staycation brings a new idea to revitalize the sagging heritage business.

Conclusions

From inception through the latest wave, COVID-19 continues to exacerbate constraints to travel. Staycation is increasingly recognized as an important alternative for leisure and travels. It helps in the recovery of a tourism industry beset by pandemic restrictions and decreased willingness to travel due to COVID outbreak. Using extant literature and data collection, staycation pivoted to a safer, closer-to-home concept has quickly become simulacra offering a short break from everyday life. For tourist destination like Macau where the majority of staycationers work in the tourism and hospitality sectors, it focuses on the emotional recovery to rest and recharge their batteries as an increasing number of local employees were required to take a furlough and no-paid leave during the pandemic. The participation in staycation offers many of the same mental and physical benefits as a traditional vacation, but with less stress and a lower price tag. The peace of mind, pleasant surprises, and hedonic happiness generated from three separate routes form a place of memory and sense of belonging in trying times.

From heritage planning perspective, staycation featuring a variety of industry collaborations and in partnership with major attractions succors to revive economic activities. Heritage tourism has offered an opportunity to look at its destination in unique lights and appreciate what locals have in their own backyard. Transforming heritage tourism to staycation creates travel bubbles that attract more local visitors exploring their areas and regions within easy walking and driving distances. However, while conventional heritage visitors want to experience something different that is preferably "educational", the staycation era shows that visitors tend to emphasize on peace of mind, inner calm, and nostalgic locations, rather than educational component of travels.

This chapter suggests that staycation is a catalyst of change and should be part of the solution for heritage tourism planners and marketers to formulate future strategies in designing and promoting their tourism products. It provides a new perspective to explore and rediscover their hometown and heritage displays. Innovative programs in the context of heritage tourism, such as those featuring storytelling that espouses visitors to relate, question, compare, or think their destinations, thereby activating a personalized authenticity. This strategy is especially important for heritage-based staycation where storytelling is an integral part of marketing tools for new places gained tremendous popularity. From a managerial perspective, the importance of staycations delivers considerable potential in this emerging strand of travel industry, finding the fit between community and travels, and serving an incentive to transform heritage tourism.

References

Alejziak, W. (2013). Tourist activity inhibitors. *International Journal of Culture, Tourism and Hospitality Research*, *7*(1), 11–27.

Alexander, A. C., Lee, K.-H., & Kim, D.-Y. (2011). Determinants of visitor's overnight stay in local food festivals: An exploration of staycation concept and its relation to the origin of visitors. *Proceedings of the 16th Annual Graduate Education & Graduate Student Conference in Hospitality & Tourism.*

Anderson, B. (1991). *Imagined communities*. Verso.

Andriotis, K. (2018). *Degrowth in tourism: Conceptual, theoretical and philosophical issues*. CABI.

Ashbullby, K. J., Pahl, S., Webley, P., & White, M. P. (2013). The beach as a setting for families' health promotion: A qualitative study with parents and children living in coastal regions in Southwest England. *Health & Place*, *23*, 138–147.

Bafadhal, A. (2021). Staycation during Covid-19 pandemic with virtual tourism: Teletourist toward experience in cultural heritage destination. *Journal of Tourism and Development Studies*, *9*(2), 87–93.

Besson, A. (2017). Everyday aesthetics on staycation as a pathway to restoration. *International Journal of Humanities and Cultural Studies*, *4*(2), 34–52.

Byrne, B. M. (2013). *Structural equation modeling with EQS: Basic concepts, applications, and programming*. Routledge.

Certeau, M. (2011). *The practice of everyday life*. University of California Press.

Chen, C. F., & Chen, F. S. (2010). Experience quality, perceived value, satisfaction and behavioral intentions for heritage tourists. *Tourism Management*, *31*(1), 29–35.

Chhabra, D. (2021). *Resilience, authenticity and digital heritage tourism*. Routledge.

Cho, K. S., Um, S. H., & Lee, T. J. (2016). Perceived restorativeness of visits to cultural heritage sites. *Asia Pacific Journal of Tourism Research*, *21*(9), 1046–1069.

Gonçalves-Sá, J. (2020). In the fight against the new coronavirus outbreak, we must also struggle with human bias. *Nature Medicine*, *26*, 305.

Hair, J. F., Black, W. C., Babin, B. J., & Anderson, R. E. (2010). In the fight against the new coronavirus outbreak, we must also struggle with human bias. *Multivariate Data Analysis Seventh Edition Prentice Hall* (New Jersey: Prentice Hall).

Henseler, J., Ringle, C. M., & Sarstedt, M. (2015). A new criterion for assessing discriminant validity in variance-based structural equation modeling. *Journal of the Academy of Marketing Science*, *43*(1), 115–135.

Henseler, J., & Sarstedt, M. (2013). Goodness-of-fit indices for partial least squares path modeling. *Computational Statistics*, *28*(2), 565–580.

Higgins-Desbiolles, F. (2020). Socialising tourism for social and ecological justice after COVID-19. *Tourism Geographies*, *22*(3), 610–623.

James, B., Matthew, B., & Barbara, T. (2017). A critical appraisal of individualised marketing and travel blending interventions in Queensland and Western Australia from 1986–2011. *Travel Behaviour and Society*, *8*, 1–13.

Janssen, J., Luiten, E., Renes, H., & Rouwendal, J. (2014). Heritage planning and spatial development in the Netherlands: Changing policies and perspectives. *International Journal of Heritage Studies*, *20*(1), 1–21.

Jeuring, J., & Haartsen, T. (2017). The challenge of proximity: The (un) attractiveness of near-home tourism destination. *Tourism Geographies*, *19*(1), 118–141.

Kalman, H. (2014). *Heritage planning* (1st ed.). Routledge.

Kaplan, S. (1995). The restorative benefits of nature: Toward an integrative framework. *Journal of Environmental Psychology*, *15*(3), 169–182.

Ko, D. (2013). The psychological effect of nature tourism. *Korean Psychological Association Journal*, *14*(3), 445–470.

Koselleck, R. (2006). Crisis (trans. Richter, M. W.). *Journal of the History of Ideas*, *67*(2), 357–400.

Larsen, J. (2008). De-exoticizing tourist travel: Everyday life and sociality on the move. *Leisure Studies*, *27*(1), 21–34.

Laumann, K., Gärling, T., & Stormark, K. M. (2001). Rating scale measures of restorative components of environments. *Journal of Environmental Psychology*, *21*(1), 31–44.

Lee, S., & Hyun, M. (2003). The factorial structure of restorative environment perception scale. *Journal of Korean Psychological Association*, *8*(2), 229–241.

Lin, Z., Wong, I., Kou, I., & Zhwn, X. (2021). Inducing wellbeing through staycation programs in the midst of the Covid-19 crisis. *Tourism Management Perspectives*. https://doi.org/10.1016/j.tmp.2021.100907

Lou, L., & Xie, P. (2018). Reawakening Macanese identity: The impact of Chinese outbound tourism. *Journal of Policy Research in Tourism Leisure and Events*, *10*(2), 253–267.

MacKerron, G., & Mourato, S. (2013). Happiness is greater in natural environments. *Global Environmental Change*, *23*(5), 992–1000.

Marselle, M. R., Irvine, K. N., Lorenzo-Arribas, A., & Warber, S. L. (2016). Does perceived restorativeness mediate the effects of perceived biodiversity and perceived naturalness on emotional well-being following group walks in nature? *Journal of Environmental Psychology*, *46*, 217–232.

Mason, K. (2004). Sound and meaning in aboriginal tourism. *Annals of Tourism Research*, *31*(4), 837–854.

Mehmetoglu, M. (2007). Typologising nature-based tourists by activity—Theoretical and practical implications. *Tourism Management*, *28*(3), 651–660.

MGTO (2020). Macao Government Tourism Office. https://www.macaotourism.gov.mo/zh-hant/

Molz, J. (2009). Representing pace in tourism mobilities: Staycations, slow travel and the amazing race. *Journal of Tourism and Cultural Change*, *7*(4), 270–286.

Moon, H., & Chan, H. (2022). Millennials' staycation experience during the Covid-19 era: Mixture of fantasy and reality. *International Journal of Contemporary Hospitality Management*, *34*(7), 2620–2639.

Moura, N. (2020). *Local tours scheme injected some MOP56 million to economy*. Macau Business. https://www.macaubusiness.com/local-tours-scheme-injected-some-mop56-mln-to-economy-govt/

Patiwael, P. R., Groote, P., & Vanclay, F. (2019). Improving heritage impact assessment: An analytical critique of the ICOMOS guidelines. *International Journal of Heritage Studies*, *25*(4), 333–347.

Pyke, S., Hartwell, H., Blake, A., & Hemingway, A. (2016). Exploring well-being as a tourism product resource. *Tourism Management*, *55*, 94–105.

Salazar, N. (2021). Existential vs. essential mobilities: Insights from before, during and after a crisis. *Mobilities*, *16*(1), 20–34.

Sharma, S. (2009). The great American staycation and the risk of stillness. *M/C Journal*, *12*(1), 13–25.

Timothy, D. J., & Boyd, S. W. (2003). *Heritage tourism*. Pearson Education.

Van Assche, K., & Duineveld, M. (2013). The good, the bad and the self-referential: Heritage planning and the productivity of difference. *International Journal of Heritage Studies*, *19*(1), 1–15.

Waterton, E., & Watson, S. (2013). Framing theory: Towards a critical imagination in heritage studies. *International Journal of Heritage Studies*, *19*(6), 546–561.

Wixon, M. (2009). *The great American staycation: How to make a vacation at home fun for the whole family (and your wallet!)*. Simon and Schuster.

Xie, P. (2015). A life cycle model of industrial heritage development. *Annals of Tourism Research*, *55*, 141–154.

Xing, L. (2022). Planning a staycation? You can claim a tax credit if you travel in Ontario this year. Canadian Broadcasting Company (January 2). https://www.cbc.ca/news/canada/toronto/tourism-tax-credit-ford-government-1.6278799

Yan, Q., Shen, H., & Hu, Y. (2022). "A home away from hem": Exploring and assessing hotel staycation as the new normal in the Covid-19 era. *International Journal of Contemporary Hospitality Management*, *34*(4), 1607–1628.

4

TRANSFORMATIVE TEA HERITAGE TOURISM IN NORTHERN THAILAND

Lee Jolliffe, Pairach Piboonrungroj, and Jirawan Chaikor

Introduction

Tea heritage-related tourism in tea-producing regions has the potential to enhance the health, well-being and depth of experience for tourists and at the same time improve the livelihoods of tea farmers and villagers (Jolliffe, 2022a). The consumption of tea has health benefits and tea-growing areas visited are typically rural and green, providing mentally calming environments for co-creation between guests and hosts. For the tea business enterprises located in growing areas, tea heritage tourism thorough experiential activities such as tours, courses, workshops, gastronomy and cuisine offer the opportunity for diversification of tea business offerings and therefor increased sustainability of the environment, workforce and tea communities. For the farmers and workers in tea villages, tourism provides the opportunity to augment their agricultural incomes through tourism, providing employment opportunities in rural areas (Jolliffe, 2022b).

This chapter identifies forms of transformative tea heritage tourism being developed by tea companies and enterprises in the tea-growing region of Northern Thailand (Chiang Mai and Chiang Mai provinces). The area is the home of the native camellia sinesis variety assamica and the imported camellia sinensis variety sinensis tea plants, both suitable for the cultivation and through processing the production of many different types of tea, including white, green, black, oolong and scented teas. The presence of the old wild assamica trees in the forest is a particular feature of the region. This forms a resource both for the sustainable production of wild tea from the offshoots of these old trees and for the touristic experiences with visitation to the trees in protected areas. This tea plucked from the forest versus

DOI: 10.4324/9781003346197-4

being cultivated is now being sustainably produced and marketed as forest-friendly tea.

Literature Review

The brief literature review that provides a context for this chapter has a focus on sustainable marketing in relation to heritage tourism, and the transformative potential of heritage tourism and tea heritage tourism.

Sustainable Marketing

As the resources for heritage tourism are finite, there is a need to develop strategic marketing plans to ensure cultural and heritage sustainability of host communities, planning in concert with locals. Marketing strategies can also encourage responsible consumption and mindfulness among visitors toward the host community and the natural landscapes of the places visited (Chhabra, 2010). As noted by Chhabra (2010), sustainable marketing is a complex concept, combining elements of sustainability, marketing and heritage. Font and McCabe (2017) note that sustainability marketing can employ marketing skills and techniques by understanding market needs and designing more sustainable products while also identifying communication methods to bring behavioral change among visitors. Studying museums as heritage institutions, Chhabra (2009) proposed a sustainable marketing framework for heritage tourism. The study offered suggestions on how the marketing plans of museums could include sustainable elements such as environmental analysis and level of local community involvement, thus potentially contributing to a sustainable paradigm in heritage tourism (Chhabra, 2009).

Transformative Heritage Tourism

Heritage tourism can be transformative, personally for both visitors and host communities. The heritage transformation process can be seen from the personal, local and regional perspectives as documented by Pranskūnienė and Zabulionienė (2023). Principles of sustainable marketing of heritage tourism noted above can be employed to impact both the personal transformative visitor experience and the preservation of the cultural heritage of local and regional host communities. Tourism is therefore employed by many destinations to enable development and achieve sustainable economic growth, thus transforming communities (Font & McCabe, 2017). Pranskūnienė and Zabulionienė (2023) note that the future of heritage transformations will be influenced by the interconnectivity of individual, local and regional perspectives creating interwoven tourism values and experiences.

Tea Heritage Tourism

Tourism related to tea has been defined as being motivated by an interest in the history, traditions and consumption of tea (Jolliffe, 2007). The niche of tea tourism has particular relevance to destinations producing tea, as tea estates and gardens can use tourism to sustainably diversify their business operations with the added goal of improving local livelihoods through increased employment and supplemental revenue streams (Su et al., 2019). Studies in China have identified that tea tourism may benefit local tea farmers (Chee-Beng & Yuling, 2010) as the tradition of the old tea industry is reinvented for contemporary tourism.

Integrating tea and tourism can therefore incorporate a sustainable livelihoods approach for the benefit of local communities, as has been noted in China (Su et al., 2019). Tea, when merged with tourism, offers unique opportunities for exhibiting tea traditions that will appeal to visitors. Beyond the beverage of tea, the consumption of tea as a part of a regions heritage cuisine and related tea gastronomy also offers a unique opportunity for development for tourism, providing meaningful experiences for visitors while conserving and preserving local food and beverage traditions.

Methods

Beyond a related literature review, the chapter is based on field work by the authors (participant observation and interviews) in both 2018 and 2023. Field work was thus completed both before and after the global COVID-19 pandemic that influenced tourism arrivals and product development in the region. The authors had previously examined the tea tourist experience in Northern Thailand (Jolliffe & Piboonrungroj, 2020) and worked on a development project with one of the tea villages.

An inventory of the tea enterprises branching out from tea as an agricultural production into offering various tea heritage-related experiences is based on the knowledge of the authors, as well as industry sources. For example, several of the authors attended the World Tea Expo, in Chiang Mai, Thailand, early in 2023 and also visited with a number of members of the tea industry. Related internet sources have also been consulted. Beyond the inventory of enterprises and their products, a SWOT analysis was conducted on tea heritage tourism in Northern Thailand. The SWOT analysis method, consisting of internal strengths and weaknesses and external opportunities and threats, has typically been used in tourism planning.

Tea in Northern Thailand

A number of tea traditions coexist in Northern Thailand. Native assamica tea leaves found in the forests of the region were traditionally harvested for the production of miang. This is a traditional fermented tea food that has been part of the foodways of the Lanna people for several hundred years, eaten after meals or served at ceremonial celebrations (Khanongnuch et al., 2017). Miang production is now transitioning to what is termed "forest friendly tea" picked from the forest. For example, tea from the forest is now harvested for and processed by Monsoon Tea, who also use the traditional miang as an ingredient in the food served in the restaurant at their Chiang Mai shop and headquarters (Monsoon Tea, n.d.).

Many of the former Kuomintang (KMT) soldiers of Yunnanese origin who migrated down from China to Burma (now Myanmar) transitioned from their former military occupations with the KMT and as mercenaries with the Thai government supporting the fight against communism and involvement as opium and heroin traders and smugglers to working in tea. They became farmers and traders, encouraged by technical support for the industry and development aid from the Republic of China (Taiwan) (Hung & Baird, 2017). Today many of the descendants of the KMT soldiers continue to be involved with tea. Due in part to the technical support from Taiwan, the tea grown is mostly camellia sinesis sinensis, and the processing is in the Chinese style, so, for example, besides green and red (black tea), oolong and puer are also produced from the leaf. From 1969, the Thai government established Thai Royal projects in the northern region with the aim of replacing opium production with cash crop plantations that included tea (Latt & Roth, 2015).

Many of the hill tribe people in the region (for example, the Akha, Lahu, Hmong and Lisu) are now involved with tea as tea farmers, workers and tea enterprise operators. As mentioned above, the transition from the production of opium to cash crops that included tea was encouraged by the Thai government and supported by the Thai Royal projects.

A blended Thai tea product has been developed by the ChaTraMue (Number One Tea Brand) that was established in 1945. Imported red (black tea) was blended to make Thai Milk Tea and Thai Black Tea, both served with ice to meet the needs of the Thai people living in a tropical climate. The company developing this beverage had been involved in tea, importing from China, for many years prior to the establishment of the brand, and this forms part of the tea heritage of the region.

A few tea attractions have developed, centered around tea gardens. For example, near Chiang Rai, Choui Fong Tea Plantation established in 1977 now operates a tea café and shop and has an exhibit about the tea located in their tea gallery, in a welcoming and architecturally designed building opened in 2015 (Choui Fong, n.d.). At the tea café and shop, demonstrations on preparing tea are also given. Also, Araksa Tea Garden has been recently developed as a tea-producing garden offering a variety of experiences, including day tours and workshops (Araksa Tea, n.d.). While it is not directly marketed as a tea attraction, Singha Park near Chiang Rai includes tea plantations

TABLE 4.1 Heritage Tea Traditions of Northern Thailand

Group	*Type of Tea*	*Product*	*Examples/ Contemporary Companies*
Hill Tribes, historic and contemporary	Assamica	Miang, wild forest tea	Monsoon Tea
Hill Tribes, contemporary	Assamica, Sinensis	Wild tea, white tea, red tea	Ahpa Tea
Former KMT, Taiwanese influence	Sinensis, Assamica	Oolong, Green Tea, Red (black) Tea, Puer	Wang Put Tang
Thai Estates/Farmers	Assamica, Sinensis	White, Green and Red (black) Teas, blended and flavored teas	Raming, Wawee
Thai Beverage Companies	Assamica	Thai Tea (Product)	ChaTraMue (Number One Tea Brand)

growing both assamica and sinensis as a key element (Singha Park, n.d.). Here, the tea plantation landscapes provide a backdrop for activities that include climbing, cycling and hiking. A tea blending workshop activity is also provided for visitors.

Therefore, various geographical and political circumstances in the region have led to the differing traditions and practices that surround the beverage of tea, offering an unusual resource for diverse tea experiences for visitors and varied tourism developments around tea. This is significant as other tea heritage tourism destinations in the world may only offer one type of tea tradition (Table 4.1).

Tea Heritage Tourism in Northern Thailand

This section highlights some of the formal tea heritage tourism projects operated by tea companies in Northern Thailand that are connected with communities, offering opportunities for transformative heritage tourism, that engages visitors and benefits communities. First, the operations and tea tourism-related products of some tea companies are highlighted (Table 4.2).

Ahpa Tea

Ahpa Tea is Lahu (hill tribe) owned and operated with benefits going back to the local Lahu hill tribe community at Doi Pun Muen in the form of employment. Here, both community-based tourism (CBT) and community service volunteer programs contribute back to the rural village by assisting with the development of vital infrastructure, including the improvement of roads and

TABLE 4.2 Tea Company Operations and Tea Tourism Products

Company	*Location*	*Product*	*Benefits*
Ahpa Tea	Fang, Chiang Mai	Tea Tea Education Tea Tours Tea House Tea Themed Cuisine and Restaurant Tea-Related Accommodation	Local Lahu community livelihoods Responsible Use of Village and Forest Lands
Araksa Tea Garden	Chiang Mai Province	Tea Tea Tours Tea Café	Employment of local hill tribe population Local purchasing
Akha Tea 1x2	Chiang Rai Province	Tea Tea Tour Map (Toyota Foundation Project) Tea Gastronomy (Toyota Foundation Project)	Diversification of tea and tourism products Livelihoods of local Akha tea workers Education of young residents on tea potential
Raming Tea/ Asian School of Tea		Tea Tea Education Tea Courses and Tours Raming Tea House Siam Celadon	Livelihoods of Akha and hill tribe communities living around tea estates Education for Northern Thailand tea community

development of the tea nursery. The tea heritage here dates back to the transition from opium production to tea in the region, and this story is part of the experiential tourism product, through an exhibit at the Lahu owned and operated Phumanee Lahu Home Hotel, located in nearby Fang District. Thus, through tea production, the operation of a hotel and CBT, local Lahu people are employed and contributions are not only to the improved livelihoods but also to the village through a percentage of CBT sales. During the COVID-19 period through relief funds and contributions, a tea house was built in the village, now operating as a learning center for tea education and hosting sessions on community enterprise and the circular economy (see Figure 4.1).

Araksa Tea Garden

Araksa's Organic Garden is sustainably managed, in terms of both the cultivation and production of the tea and the social responsibility of the local community. The tea garden is open to visitors for tours offering both on-site tea tours and tea workshops. Araksa both employs local people and teaches them new skills. Their employees are from the local hill tribes (Akha, Hmong and Lisu) and local Thais; Araksa states that "we believe people are

FIGURE 4.1 Tea Learning Center

Source: Authors.

the bedrock of a sustainable community" (Araksa, n.d.). Wherever possible, Araksa also purchases locally.

Akha Tea 1x2

The tea village of Phaya Phrai was partnered with Umegashima in Japan as part of the Toyota Foundation-funded project "Revitalizing Tea Industry Community through Gastronomical Tea Tourism". The project developed tea gastronomy at both locations, including recipes and tour itineraries (see Figure 4.2). Akha Tea 1x2 was a major participant in the project that supported new tourism businesses and also served to encourage the development of new products by young tea farmers and entrepreneurs in the community. Villagers also collaborated with project leaders to build an image of the Akha village as a gastronomical tea tourism destination.

Raming Tea/Asian School of Tea

Raming Tea Estate operates the oldest cultivated tea garden in the region, established in the 1940s. Recently, Raming has established the Asian School of Tea, Thailand (ASTT), which is located at its Raming Tea House, in a

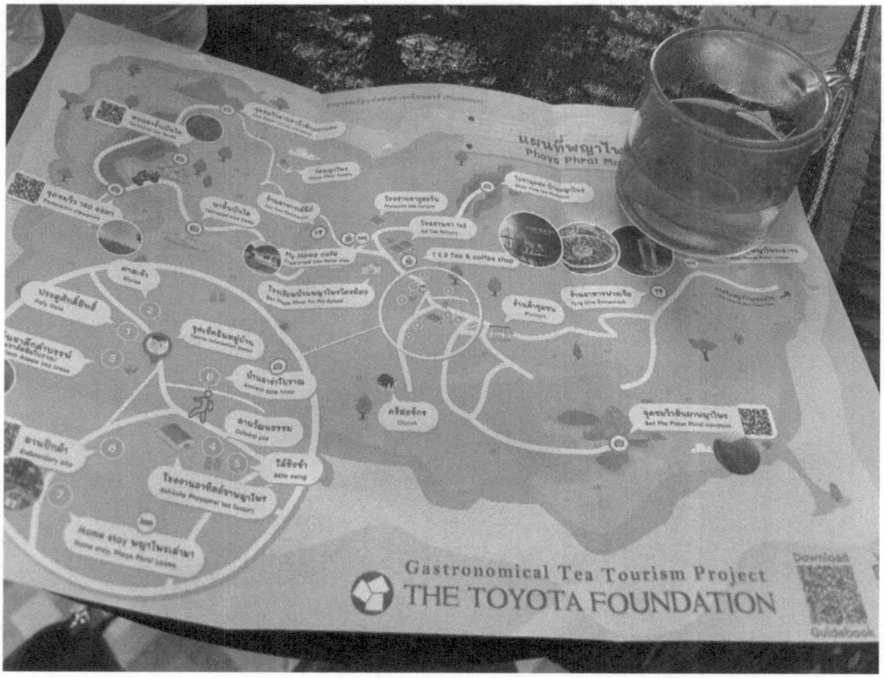

FIGURE 4.2 Itinerary for Gastronomical Tea Tourism at Phaya Phrai, Thailand

Source: Authors.

restored colonial building in Chiang Mai. The tea school is now offering a comprehensive menu of tea education, including tea sommelier certification and tea blending and pairing classes. These hands on experiences are designed to appeal to the general public and specialists, thus raising the awareness of tea produced in Thailand, while contributing to developing the tea heritage tourism knowledge base in the region. They are also able to offer special sessions to tie in with tourism focused events in the city, such as Chiang Mai Blooms, an annual flower festival, for which the ASTT offered a hands-on workshop on blending tea with flowers.

Analysis

The inventory of tea enterprises offering tourism experiences shows a gradual transition from just producing tea to also offering tea experiences, contributing to sustainability, enhancing the income of local workers and thus potentially improving livelihoods. This transition began before the COVID-19 pandemic but has been accelerated in the post-pandemic period. For example, the Araksa Tea project established in 2018 is now fully operational, offering a variety of workshops and experiences within its sustainability mission listed on its web site (Araksa Tea, n.d.). The newly established ASTT with programs operating since 2022 has emerged post-pandemic as a major influencer in terms of tea education in the region, attracting both locals and visitors to participate in learning about tea. As a diversification of Raming Tea, this also indirectly supports the livelihoods of the ethnic hill tribe communities surrounding the estate. Operating from a restored heritage building in Chiang Mai, the Raming Tea House and the ASTT are also contributing to heritage appreciation and conservation.

The variety of experiences offered to visitors are potentially transformative for the tea tourist participants. Reflective of the differing tea traditions of Northern Thailand, the heritage tourism products offered are also related to the various food and beverage ways of the people of the region. There is some evidence of tea-based gastronomy, incorporating tea into food menus. Based on the inventory (Table 4.2), a SWOT analysis was conducted on the current state (2023) of tea heritage tourism in the region (Table 4.3).

TABLE 4.3 SWOT Analysis of Tea Heritage Tourism in Northern Thailand

Strengths	*Opportunities*
• Tea history of region • Tea landscapes, gardens and villages • Tea heritage and various traditions • Communication and partnerships between regional tea tourism players • Strong connection of tea enterprises to local communities • Tourists seeking authenticity through unique food and beverage experiences • Tour operator infrastructure exists • Return of tourists to Thailand (post-pandemic)	• Destination policy development to support expansion of tea tourism • Developing visitor base through membership clubs and CSA • Further development of tea gastronomy element in all products • Expansion of tea tour and tea education products • Development of sustainability missions by tea enterprises and tea tourism product managers
Weaknesses	*Threats*
• Rural locations of some tea locations, distance, road and travel conditions • Distance of some tea farms and related enterprises from airports where international tourists arrive (Chiang Mai and Chiang Rai) • Limited accommodation and food services in some rural areas • Training needed on all aspects of tea tourism, including sustainable marketing, experiential product development, etc.	• Lack of tourist awareness of Northern Thailand as a tea destination • Competition from other Asian tea destination countries and tea regions (China, Japan, Taiwan, etc.) • Seasonality of tea production and related experiences • Environmental issues, i.e., the "burning season" threatening visitation and tea community health Climate change impacts to growing season and weather

Strengths

There are many strengths in terms of tea heritage tourism in Northern Thailand. There is a rich tea history and tea-related heritage to be found amid the landscape of tea gardens and villages in the region. Since there are varied tea traditions, visitors can have a broad variety of different experiences with tea and local cultures and cuisines. The strong connection of tea to local communities and ethnic hill tribes is of particular note at a time when tourists are seeking authenticity in their experiences. There is good communication in the tea industry and some partnerships exist between different players in tea tourism in the region. There is a strong tour operator infrastructure which is significant to developing experiences in the rural tea-growing region and communities, where transportation and other logistics need to be handled by professionals. Finally, the emergence of tea tourism in post-COVID-19 has seemed to renew and regenerate the tea industry who is now more actively involved in providing experiences for visitors, while being aware of the potential to transform and support rural communities through such tourism.

Weaknesses

While the rural locations of some tea farms and enterprises are an advantage in terms of authentic settings, it may also be a weakness in terms of the distance, road and travel conditions, especially in relationship to airports where international tourists arrive (Chiang Mai and Chiang Rai). In these rural areas, there may be limited accommodation options, thus limiting length of stay, and food services may not be extensive. As tea tourism is a new product for many entrepreneurs in the tea industry, it would be beneficial to have training for local operators on all aspects of tea and related tourism, including sustainable marketing and experiential product development.

Opportunities

There are many opportunities for expansion and ongoing development of the heritage tea tourism product in Northern Thailand. First, as a foundation destination policy, development could support the expansion of tea tourism. The tea enterprises could further develop their visitor base through membership clubs with events and subscriptions. Such efforts could use the community-supported agriculture (CSA) model, as has been done by Obobu Tea Farms in Japan and Renegade Tea Estate in Georgia. Membership or subscription programs give these enterprises a regular source of income to support their work and development, and a dedicated base of tea lovers to promote and share their tea with others. There is also an opportunity for the further development of tea gastronomy element in all products as is evidenced by the Toyota Foundation tea gastronomy project efforts in Phaya Phrai, Thailand. So, all existing tea tour and tea education products could be expanded and further developed for increased variety of experiences and capacity. Exchange of knowledge about tea and related tourism products and developments is also important. This could be accomplished by the implementation of sustainability missions by tea enterprises and tea tourism product managers.

Threats

There may be a lack of tourist awareness of Northern Thailand as a tea destination versus some of the better-known Asian tea destination countries and tea regions (China, Japan, Taiwan, etc.) who also may be competitors in the market for attracting tea heritage tourists. There are some limitations to the ongoing development of tea heritage tourism in the region. To some extent, the seasonality of tea production limits the quality of the related experiences that visitors can have. While tea experiences can be had year round at the producing locations, it would be most interesting to visit during the season when tea is being harvested and processed (from the middle of February to October), especially if visitors are seeking deep knowledge of tea. Also, there are several environmental challenges, both from climate change threatening tea production and from the "burning season" threatening visitation and tea community health. During this period in early spring (approximately mid-February to mid-April) when farmers in the north and in neighboring countries (Myanmar and Laos) burn off their fields, there is a health-threatening decrease in air quality, and there may also be forest fires. This burning period coincides with the start of the tea-picking and -producing season when dedicated tea tourists would want to visit the region to experience the start of the tea harvest season.

The first threat of the lack of awareness of Northern Thailand as a tea destination may be overcome through marketing and promotion of Thai tea and related experiences, especially in the regional Asian market and internationally. Tea enterprise operators can work around the issue of seasonality of tea production by creating small-scale off-season experiences where visitors can still pluck tea leaves and produce a small amount that they can take home as a souvenir. The environmental challenges are more difficult to overcome and will require preventative planning and adaptation.

Conclusion

Tea heritage tourism in Northern Thailand is definitely in a developmental stage, encouraged by the post-COVID travel and tourism trend and reflective of the growing maturity of the regional tea industry region. It is notable that new experiential offerings for tea heritage tourism have emerged post-COVID-19. These products are increasing opportunities at the personal level for transformative experiences with tea and also contributing to the sustainable transformation of communities on both a local and a regional basis.

There are certainly challenges to be overcome in terms of both logistics and capacity, with tea experiential products being located in rural areas sometimes having limited accommodation that would encourage longer stays, thus positively impacting the economics of local communities through tourism. Northern Thailand also faces competition from other more developed tea destinations in Asia, although there is also the opportunity to learn from these places with more developed tea tourism offerings, and there is some evidence of related exchanges of knowledge. The environmental challenges are more difficult and will require adaptation of both the tea industry and related heritage tourism initiatives.

Acknowledgment

The author would like to thank the tea industry community of Northern Thailand for access to information that contributed to this work, in particular: Piyaporn Chueamchaitrakun of Mah Fah Luang University, Kenneth Rimdahl of Monsoon Tea, Nanzii Cino and Orapakk Ruttphatai of the Asian School of Tea, Thailand, Leo Shevchenko of the Tea Forest Project and Mr. Udom of Udomporn Tours in Chiang Mai.

References

Araksa Tea. (n.d.). Available from https://www.araksatea.com/ Accessed April 23, 2023.

ChaTraMue. (n.d.). Available from https://www.cha-thai.com/en/ Accessed April 19, 2023.

Chee-Beng, T., & Yuling, D. (2010). The promotion of tea in South China: Re-inventing tradition in an old industry. *Food and Foodways*, *18*(3), 121–144.

Chhabra, D. (2009). Proposing a sustainable marketing framework for heritage tourism. *Journal of Sustainable Tourism*, *17*(3), 303–320.

Chhabra, D. (2010). *Sustainable marketing of cultural and heritage tourism.* Routledge.

Choui Fong. (n.d.). Available from https://www.chouifongtea.com/, Accessed April 23, 2023.

Font, X., & McCabe, S. (2017). Sustainability and marketing in tourism: Its contexts, paradoxes, approaches, challenges and potential. *Journal of Sustainable Tourism*, *25*(7), 869–883.

Hung, P. Y., & Baird, I. G. (2017). From soldiers to farmers: The political geography of Chinese Kuomintang territorialization in northern Thailand. *Political Geography*, *58*, 1–13.

Joliffe, L. (2007). Tea Tourism: Global Trends and Development. *In Joliffe*, L. and Zhuang, P. (2007). "Connecting Tea and Tourism", Tea-Tourism: Global Trends and Development, Joliffe, L. (Ed.). (pp. 3–20). Channel View Publications, Clevedon.

Jolliffe, L. (2022a). Tea heritage and tourism as sustainable development in the Eastern Himalaya. In G. P. Nyaupane, & D. J. Timothy (Eds.), *Tourism and development in the Himalaya* (pp. 225–241). Routledge.

Jolliffe, L. (2022b). Tea tourism. In D. Buhalis (Ed.), *Encyclopedia of tourism management and marketing* (pp. 338–340). Edward Elgar Publishing.

Jolliffe, L., & Piboonrungroj, P. (2020). The role of themes and stories in tourism experiences. In S. Dixit (Ed.), *The Routledge handbook of tourism experience management and marketing* (pp. 218–228). Routledge.

Khanongnuch, C., Unban, K., Kanpiengjai, A., & Saenjum, C. (2017). Recent research advances and ethno-botanical history of miang, a traditional fermented tea (Camellia sinensis var. assamica) of northern Thailand. *Journal of Ethnic Foods*, *4*(3), 135–144.

Latt, S., & Roth, R. (2015). Agrarian change and ethnic politics: Restructuring of Hmong and Shan labour and agricultural production in Northern Thailand. *Journal of Agrarian Change*, *15*(2), 220–238.

Monsoon Tea. (n.d.). Available from https://monsoontea.co.th/ Accessed April 20, 2023.

Pranskūnienė, R., & Zabulionienė, E. (2023). Towards heritage transformation perspectives. *Sustainability*, *15*(7), 6135.

Singha Park (n.d.). Available from https://www.singhapark.com/ Accessed May 4, 2023.

Su, M. M., Wall, G., & Wang, Y. (2019). Integrating tea and tourism: A sustainable livelihoods approach. *Journal of Sustainable Tourism*, *27*(10), 1591–1608.

5

SOCIAL TAGGING MUSEUM COLLECTIONS

Public Engagement or an Abrogation Act

Taylor Pintel and Deepak Chhabra

Introduction

The term museum was used in the Roman times to refer to venues hosting discussions on various philosophies (Lewis, 2019). Later (from 15th to 17th century), in Europe, the notion of museum evolved to portray art. Museums were referred to as "cabinets of curiosity" representing the aristocrats and distinguished sages (Chhabra, 2009). Only privileged people had access to these spaces. It was around the 18th and 19th centuries that the museums embraced a public-centered vision. The new museology presented the museums as "interdisciplinary, decolonized ……." vehicles of social change and a locus of community participation and symbolic exchanges" (Gonsales, 2021, p. 320). Contemporary discourse continues to redefine the museums as concerns have been raised with regard to the inability of the museums to strategize their duties and address issues relevant to the 21st century. Emphasizing on the museum role to serve the society, ICOM redefined the purpose of museums:

> A museum is a non-profit, permanent institution in the service of society and its development, open to the public, which acquires, conserves, researches, communicates and exhibits the tangible and intangible heritage of humanity and its environment for the purposes of education, study and enjoyment.
>
> *(ICOM, 2007 cited in Gonsales (2021))*

The above definition by ICOM expects the museums to perform a multitude of tasks. A more recent definition by ICOM describes museums as

DOI: 10.4324/9781003346197-5

"democratizing, inclusive and polyphonic spaces for critical dialogue about the pasts and futures" [...] "to contribute to human dignity and social justice, global equality, and planetary well-being" (Marshall, 2020). In the contemporary times, museums have to compete with other educational, leisure, recreation, and entertainment enterprises to draw visitor attention, time, and spending (Komarac et al., 2017). In addition, they are entrusted with the responsibility of taking the well-being of the society into consideration. Therefore, it has become critical for museums to devise innovative strategies to engage visitors using a variety of experiential techniques. Particularly, the post-pandemic times call for museums to integrate innovative technology tools such as augmented reality (AR), virtual reality, and other interactive platforms to enable engaging and immersive visitor experiences (Loureiro, 2020). Traditionally, the digital channels were initially tapped by museums to support and enhance the physical visits. However, they have become a significant part of the museum–public interaction today and deliver special online services such as knowledge of the museum collections/exhibits, participative games, and online educational modules (Hume & Mills, 2011). According to Kabassi (2017) and Chhabra (2021), strategically designed websites and mobile applications hold potential to cultivate interest in museums, leading to more visitors and development of engaging and novel experiences.

Speaking of the testing pandemic times, Gonsales writes, "as the only mode of visiting during the pandemic, the online Place is a faster and safer alternative, allowing the public to explore the museum in a photographic or 3D reconstruction and to shop at the online store, any time of the day, from anywhere in the world. Although not yet universally available, nor equally adopted by museums, the digital presence has become mandatory .... configuring the era of the post-digital museum" (2021, p. 322). The online catalogues of museums hold tremendous potential by offering an open space where the public can assist in documenting and labeling collection records. Facilitated by the progress in digitalization, Gonsales (2021) further recontextualize the mission of museums to stress on their social relevance and demonstration of valuing multiple themes and stories around museum exhibits and objects valued by the audience. Furthermore, the museum professionals today are recognizing the role and power of non-human actors (e.g., data and technology) along with the missing actors (the public) in their assistance with documenting collections. The views of multiple actors can offer new avenues of reflection.

In the past decade, with the increase in technology, social tagging and folksonomy have become popular as they hold potential to engage and direct audience attention to heritage sites (Samanta & Rath, 2021). Along similar lines, Trant and Wyman note, "museums want audiences to engage with their collections and ideas, but recognize that traditional methods of uni-directional on-line and in-gallery communications have limited access and dialog. Supporting social tagging of museum collections, and providing access based on the resulting folksonomy, opens museum collections to new interpretations that reflect visitors' perspectives rather than institutional ones" (2006, p. 1). This type of museum-visitor collaboration bypasses the hiatus between curatorial language and the common dialect of the museum visitor. According to West (2017), it has been increasingly noted that art historians and other image professionals value the use of tags (subject terms) to search for and connect with art visuals. That said, "the subject indexing of artworks and images is a subjective and time-consuming undertaking. As a result, many digital image collections are lacking comprehensive subject access points at the object level. Social tagging of digital images has been proposed as a means of improving image metadata and increasing subject access to art images" (West, 2017, p. 1).

Vander Wal (2007) describes folksonomy as a combination of two terms: (a) folk, which means people; (b) taxonomy, which means a system of classification. According to him, folksonomy is a process of personal free tagging of digital resources on social networking sites for the retrieval of individual information. Social tagging allows users to describe web resources using keywords in the form of tags and without following any standard vocabulary. Social tagging has led to the establishment of a completely new vocabulary that can be used to communicate via the Internet. In order to keep up with the technological progress in the contemporary times, museums are increasingly implementing technology to influence and communicate their collections to the public. Social taggers, whether art experts or not, are able to publish their own interpretations of art collections. Social tagging offers a way for people to give their own insights on art and assists in augmenting museum's content with collections in a manner that captures the attention of the communities. However, there are concerns that the interpretations of art done by taggers can disrupt the primary responsibility of the museums and provide viewers with unreliable information (Trant, 2006). Undeniably, the traditional role of museums has been to deliver accurate information, but with social tagging, unreliable information is being linked to museum works. However, most art museums do not include any sort of context or description in their collections because it is not required; therefore, that leaves room for anyone to provide his or her interpretation. Now, museums are using the tags supplied by visitors to gain perspective and understand the manner in which the users perceive different works of art. Furthermore, this process helps them to notice any disconnection between the users' tags and the accurate interpretation of the art piece.

Folksonomy is being used to gain the attention of a wide range of viewers, while preserving the traditional objective of museums (Loureiro et al., 2020). The motivations of social taggers, whether leisure-based or for work, are based on an individual need to promote himself or herself and other users, thereby establishing social connections. They can also be motivated by future retrieval opportunities or the autonomy to freely express their opinion (promoting a sense of autonomy). Certain museums actively support individuals who are interested in social tagging to connect with them and value their input. While other museums feel that, it diminishes the work of employees and artists. This chapter discusses the advantages and disadvantages of social tagging based on the perspectives of Generation Z.

Numerous studies have demonstrated that age plays an enormous role in shaping the motivations of people (Chung et al., 2016). The theory of generational cohorts suggests that each generation shares a unique set of values, beliefs, attitudes, and behavior that is adopted by any sort of social, political, or economic events that occurred during the pre-mature stages of their life cycle. These historical events, such as the outbreak of COVID-19, have a long-term impact on the beliefs and attitudes of people that remains consistent throughout their whole life (Chung et al., 2016). In relation to tourism, it has been demonstrated that motivation is a multidimensional construct, which recognizes an individuals' different desires within a trip. For instance, an individual on a family vacation might seek out sightseeing and relaxation within one vacation (Crompton & McKay, 1997). Generational cohorts can be broadly delineated into four categories: Baby Boomers (1945–1965); Generation X (1966–1980); Generation Y (1981–1995); and Generation Z (1996–2010) (Eger et al., 2021; Thangavel et al., 2021).

Briefly, the data for this pilot study is anchored in a purposeful sample of Generation Z student population in a controlled university setting. The overarching aim of this chapter is to offer insights on social tagging and its process (folksonomy) from a Generation Z standpoint. Specifically, the purpose is to answer the following research questions: (1) Is including folksonomic terms in museum records an abrogation of museum responsibility? (2) Does social tagging facilitate audience engagement? 3) Is social tagging a useful mechanism to engage heritage audiences belonging to the Generation Z?

Literature Review

This section offers insights on the manner museums are connecting with their audience today with the help of folksonomy. Different types of social tagging networks, noted in documented literature, are presented. This is followed by a discourse on the pros and cons of tagging. Examples of some museums indulging in this program are shared. Additionally, based on a cursory review of existing literature, insights are offered on the effects of different tagging models on users. This section closes with a discussion of Rapid Response Collection (RRC), an emerging branch of folksonomy.

The contemporary era focuses on making art museum collections accessible to communities as a way of engaging with them and obtaining insights on the way they relate to the collections so that user-friendly messages can be designed. As posited by Trant, "social tagging (the collective assignment of keywords to resources) and its resulting Folksonomy (the assemblage of concepts expressed in such a cooperatively developed system of classification) offer ways for art museums to engage with their communities and to understand what users of on-line museum collections see as important" (2006, p. 1). User tags can bridge the semantic hiatus between the curatorial dialectic and the common expressions of a museum visitor. Authored materials tend to have a very piquant museum "tone" and formal outlook (Walsh and Gribbon, 1997). In comparison, the collections databases present the attributes of the objects in a non-contextual manner and distanced from other associated works. Many giant-sized objects in museum collections can appear similar to an untrained eye and can show up as a homogeneous collage, to which the audience might not be able to relate. It has been extensively noted that specialized narratives and operational procedures might not align with perceptions and interests of the general public. In other words, for many decades, museum collections have existed but they have not been accessible or easy to comprehend (Golder & Huberman, 2006; Hudoshnyk & Krupskyi, 2023; McKenna & Patsatzi, 2005). Social tagging holds potential to resolve part of this issue by engaging visitors and obtaining insights on the manner in which they view and express their understanding of the objects. Museums of today are finding this technique appealing because the tags are an expression of a dialogue between the audience and the exhibit or artwork (Bearman & Trant, 2005). They offer a pathway "to connect directly with works of art, to own them by labeling or naming them which is a type of sensemaking. Tagging lets users assert their own connections and associations between objects in ways that reflect personal perspectives and interests" (Trant, 2006, p. 3).

Several scholars agree that bridging the semantic gap between the specialized, curatorial dialect of art history and the manner in which the public audience views this presentation and expression is an arduous task (Chhabra, 2021; Gonsales, 2021; Kabassi, 2017). Folksonomy offers an opportunity to enhance access and connect the public with facets of interest. However, Trant warns, "as collection description and interpretation is a prime professional role of the museum, ceding it to the general public seemed an abrogation of responsibility. Museums see themselves as the source for authoritative information about the objects in their care. Conversely, social tagging appears to undermine that role, letting 'just anybody' play a role in how a work is described and ultimately perceived" (2006, p. 3). For example, if unfiltered results are presented, museum works can be interpreted incorrectly and this can jeopardize a core responsibility of the museums, which is distribution of authentic/accurate information.

Several scholars support continued use of social tagging tools in art museums (Alves da Veiga, 2020; Bearman & Trant, 2005; West, 2017). The non-experts can also add to the specialized narrative of art pieces. As an example, an open-source social tagging platform is being designed at Steve Museum to support shared exploration of social tagging and folksonomy in art museums. "Steve" is a collaborative research endeavor that explores the potential for user-generated descriptions of art works to enhance access to museum collections and facilitate engagement and interactions with the cultural content. It is run by volunteers from art museums, who are motivated by a shared interest to improve access to museum collections (Steve Museum, 2008). In other words, its key goal is to facilitate audience interaction with the exhibits and it operates on a shared mission to design bonds between people and art exhibits. The museums offer a creative and novel social space to explore the prospects of social tagging and folksonomy. According to Trant (2006), the act of tagging and the assigned name – offer important acumen about how museums can relate to visitors. In their core function, "museums are institutions dedicated to serving local and global communities. Collectively, museums are charged with preserving the cultural trust. Both of these goals require an ability to connect with individuals and communities on their own terms. Social tagging and folksonomy offer a less formal, more participatory, and highly distributed way to augment museums' institutional documentation with content that reflects the perspectives and interests of their communities" (Trant, 2006, p. 17). Undeniably, art has several dimensions. However, its multiple meanings are not easy to convey or it can be arduous to identify the dimension that most relates to the audience. Tagging can assist in identifying aspects of artwork that are most engaging to the audience. Next section offers insights on platforms that are patronized frequently for social tagging.

Social Tagging Networks

Social tagging has dramatically increased in recent years (Ding et al., 2009; Gonsales, 2021; Loureiro et al., 2020). The most common social tagging networks are noted to be *Web 2.0, Delicious, Flickr, and Youtube* (Ding et al., 2009; Hekman & Vliet, n.d.; Trant, 2006; West, 2017). According to Ding et al. (2009), more and more people are utilizing social tagging applications to record online resources for future reference. With regard to *Delicious*, users are able to tag their own resources as well as those belonging to other people. Flickr is also noted to be a popular network for facilitating personal tagging (Ding et al., 2009). Taggers in this network are noted to be both nonprofessional and professional photographers and they often include the date and location of their photographs. Based on the informal nature of their tags, users of *Flickr* tag the photographs for the purpose of storing and refer back to them later for their personal *use*. Patronization of *YouTube* for social tagging has also increased (West, 2017). A large number of individuals are, therefore, shifting to these networks to categorize online content. Another plus point noted is that social tagging helps to reveal emerging trends and popular topics within a community (Trant, 2006).

Social Tagging Projects

West (2017) discussed three social tagging projects (Steve.museum, the Brooklyn Museum Posse, and Art UK's Your Paintings Tagger) which were created as test programs to explore issues regarding the social tagging of digital images in museum collections. The results of their study suggest that carefully designed tagging programs can improve access to art images when they are used as a supplement to professional indexing. Mobitags are becoming popular.

According to Cosley et al., they represent "a system to help museum visitors interact with a collection of 'open storage' exhibits, those where the museum provides little curatorial information. MobiTags integrates social tagging, art information, and a map to support navigation and collaborative curation of these open storage collections" (2009, p. 1953). Focusing on MobiTags, Cosley et al. (2009) conducted a study at the Johnson Museum of Art at Cornell University and shared that the displayed collection consisted of 51 artifacts spread on one floor of the museum. The objects consisted of ancient weights, vessels, bowls, masks, or totems. MobiTags work on an iPod Touch, and visitors are able to use the JavaScript library to offer their insights with the help of interactive navigation systems. Researchers observed the audience as they were walking around the museum and using the system. The focus was on how the participants interacted with the device and the art, and the manner in which they navigated throughout the museum. After their experience, the participants were briefly interviewed and given a questionnaire to better understand how they were utilizing MobiTags. It was noted that a solo device does not hold the capability to support three different modes of navigation. The key challenge noted was too much information displayed on a small screen which resulted in crowding the screen and discouraged people from visiting some exhibits. Because of this distraction, many users became less cognizant and responsive to their physical environment. Although Portrait mode provided users with semantic and social information about the artifact of interest and the map overview provided users with a layout of the space and it was able to guide the audience toward a particular section of the collection. It was also noted that the users found it difficult to view/enjoy the artwork, while they were trying to operate the device. Some participants were not able to visit their preferred art exhibits. Additionally, many users were reluctant to tag because they felt that they were not knowledgeable enough and needed expertise to contribute. Average duration of interacting with the MobiTags was noted to be 20 minutes. Some respondents also reported that they liked details of the artifact (made available to them) and that accessibility enabled them to study the exhibit at their own pace. It was also interesting to observe that participants mainly interacted with the objects that had the maximum number of social tags. Participants reported that they felt that the tags allowed them to view art in a new way and they were able to recognize things that they would not have noticed earlier. In addition, many participants shared that they felt connected with others when they looked at the subjective tags (such as "gorgeous" and "frightening"). The study was structured so that visitors could vote on the tags which resonated with them. This facilitated a sense of autonomy, stimulated curiosity, and social bonding with other users.

Based on the foregoing and the review of few more studies, several pros and cons of social tagging can be noted. With regard to advantages of social tagging, some studies note that the main benefit of social tagging is that it allows anyone to provide their individual interpretation and insights on a work of art. Hekman and Vliet (n.d.) reiterated this point through an experiment where a platform was developed where anyone could upload their tags for different art collections (consisting of 134 objects from the Dental Surgery collection at Utrecht University Museum, 145 artifacts from the collection of drawings from Japanese internment camps at Museon and 100 images from the Beetle's collection). A total of 3,592 tags were submitted, in one year, by both nonprofessionals *and* experts. A review of the tags revealed that the expert tags were more informative compared to those of the nonprofessionals. It was noted that after viewing, listening, and reading about the art, the audience's attitude toward the museums became more positive and visitor engagement increased. In summary, the study reported that tagging results stimulated a positive and active engagement with the artwork.

Besides enhancing audience engagement, it is also reported that social tagging holds potential to monitor trends and generate important data from users (Cosley et al., 2009; Trant, 2006). Popular tags demonstrate latest preferences and the state of mind of users In addition, how the audience relates to the information helps the museums to form strategies to cater to the market needs and exhibit artifacts that best capture the attention of the audience. In this way, the museums are able to arrange their art and showcase different works in a way that can best hold the audience attention and generate high satisfaction levels. Bearman and Trant opine that, "tagging appeals to museums partly because it embodies this philosophy: tagging represents a dialog between the viewer and the work, and the viewer and the museum" (2009, p. 3763). That is, the users are able to express their own opinions and interests based on the manner in which they relate to a work of art. It is also interesting to note that social tagging adds value to a range of existing museum collections. Viewers are frequently unable to comprehend the intentions of the artists from a brief description provided by the curators and/or artists. The tags provide immediate insight and feedback to the museums and indicate prevailing sentiments of the viewers. They also serve as direct indicators of any potential discrepancies between the audience's viewpoint and the intended message of the artist.

However, some studies note that over-reliance on public-generated social tags can prove to be damaging to the museum purpose. Several scholars argue that the information published by social taggers is not always accurate. As an instance, The Metropolitan Museum of Art conducted a study in 2004 to test this idea (Trant, 2006). Two tests were administered. The first consisted of five library professionals and five library assistants. Each was given five images and asked to reform terms on worksheets. The results of this experiment demonstrated that the library assistants provided more terms than that of the library professionals. In another test, the library assistants were given ten images and asked to place them into different categories. Consequently, it was determined that some experts find the simple descriptions of artwork frustrating and difficult to interpret. It was reported that, "users rather than professionals might provide the bulk of indexing for a document could influence cataloging practice more broadly; it definitely challenges the assumptions underlying most indexing research" (Trant, 2006, p. 16).

Shifting Role of Museums in Contemporary Times

As mentioned in the aforementioned paragraphs, while many scholars argue that social tagging defeats the purpose of museums, they also acknowledge that the museums are encouraging folksonomy because it increases audience engagement. Also, recognizing the different perspectives of individuals reflects a museum's commitment to promote inclusion and mindfulness toward diverse voices. By forming their own tag, visitors can develop a bond with the museum and with other like-minded users (Cosley et al., 2009; Trant, 2009; West, 2017). The conventional definition of museums is widely debated today because interactive museums have emerged that allow visitors to express themselves rather than gaze at the work of *artists* (Alves da Veiga, 2020). The Museum of Selfies and the Museum of Candy are both examples of modern museums that offer programs to facilitate visitor engagement. In fact, trend is toward large AR spaces (Alves da Veiga, 2020). The Museum of Everything is completely virtual and has an international outreach. Different from other museums, "The Museum of Everything is also a publisher, an archive, an activist and a leading advocate for alternative artists from the 19th, 20th, and 21st centuries" (TEME, n.d.).

In addition to the TEME example, museums have now begun incorporating their own online exhibitions that allow the audience to develop personal collections (Anderson, 2020). Museum websites are mainly focused on collections that are highly authored, linear exhibitions or un-interrupted databases and do not leave room for audience interpretation. Furthermore, the museum collections can appear similar to those who are not experts and it requires skills to distinguish between different works of art. Collections posted on such museum websites are not necessarily comprehendible to the average viewer because the descriptions are not explicit in their meaning. This shows the dilemma traditional museums encounter in the contemporary times: retaining its core function versus adapting to the needs of the contemporary audience. As pointed out earlier, the limited purpose of this traditional museology is questioned today (Tenenbaum, 2020) and museum curators are required to step outside their comfort zones to engage more with the society. The International Council of Museums (ICOM) most recently redefined the role of museums to emphasize inclusiveness and engagement with diverse groups of populations which might have diverse needs. This changing museology has paved the path for an emerging trend: RRC. The RRC of contemporary objects has the ability to expose and record political events as they are happening (Tenenbaum, 2020) and requires attention in a timely manner. Next section offers insights on the RRC phenomena.

Implementing RRC in Museums

In response to COVID-19, museums and curators who were not keen to embrace technology, were not left with much choice but to re-evaluate their practice and adjust their part in society. Part of this adjustment includes implementing RRC in a variety of ways such as through photographs or a collection of artifacts such as homemade facemasks. Museums worked to document the pandemic as it was happening. RRC is particularly designed to facilitate audience engagement while challenging museum curators to develop ways to involve the public (Tenenbaum, 2020). This can be achieved through promoting collaboration between museums and users or by simply collecting stories from local people to obtain insights on their perspectives and document them.

As an instance, the current war in Ukraine is being documented through satellite images and social media. Many people in Ukraine are sharing their stories on social networks and these platforms are engaged in documenting history (Hudoshnyk & Krupskyi, 2023). A study was conducted that focused on the immediacy of experience, integrity, and multi-platform design, organization of materials, and digitalization of content of the Ukrainian wartime museum. The Anti-Terrorist Operation Museum (ATO) and the online museum of Civilian Voices were used as test subjects for the purpose of this study. The ATO Museum and Museum of Civilian Voices collected documentaries with war chronicles, interviews, memoirs and stories on the media in order to best educate and inform young audiences and received donations of personal belongings (such as letters and ordered of their deceased family members) from several families. This helped to develop an "Open concept" where the feeling of engagement stays with the visitors the real time (Hudoshnyk & Krupskyi, 2023). Based on a detailed narrative by Hudoshnyk and Krupskyi (2023), the ATO Museum has a Hall of Memory that continues to collect and replenish old photographs with new ones. It has now become one of the most popular museums in the city of Dnipro with over a million guests and visits by 50,000 children. Nine months after the start of the war in Ukraine, museums had collected about 50,555 stories from witnesses, whereas, before the war during the time of opening in 2021, the museums had documented about 2,000 oral histories. The issue of using objects, tagging them and narrating their stories from a live dark tourism site, needs further deliberations. Since this is beyond the scope of this chapter, it will be touched upon briefly in the conclusion section from the standpoint of museum ethos regarding treatment of objects from vulnerable places.

Method

A purposeful sample of 60 students, representing Generation Z, from a university in the Southwest part of United States were surveyed over a period of three years (2021–2023) in a controlled setting. Approximately 20 students from each listed year were involved. The purpose of using the three-year timeframe was to detect change in views due to the pandemic intervention, which resulted in a time where people were heavily relying on Internet (due to various lockdown phases). To gather insights on the perspectives of social tagging, respondents were asked to work on an exercise and answer three questions: can the process of social tagging and folksonomy really promote audience engagement? Is including folksonomic terms in museum records an abrogation of museum responsibility? In other words, does social tagging "undermine the role of museums?" The third question was related to a social tagging activity. With regard to the hands-on exercise, the respondents were asked to browse through the Metropolitan Museum of Art website or the Cleveland Museum of Art website and view an "Artwork of the Day" (Today's Featured Work of Art from the Permanent Collection). They had to download a work of art and social tag it. After tagging, they were required to look for the actual interpretation of the exhibit shared by the museum and match it with their tags. Upon completion of this exercise, they were required to answer the following question: does social tagging help to facilitate audience engagement? Besides noting respondent answers in quotes, content analysis was used to identify frequently occurring themes. Two researchers coded the data and identified frequently occurring themes. Reliability test was conducted to test inter-coder reliability and it was found to be 88%.

Findings

This section shares the results of the study, both in an aggregate form and by using quotes from individual respondents, in an autonomous manner. First, responses to the audience engagement are presented. This is followed by the responses to the possible abrogation of museums question. The section closes with sharing insights based on the respondent answers after they participated in the social tagging exercise.

Audience Engagement

The respondents shared that as non-experts, in the absence of such engaging exercises, it is difficult for them to sustain their interest in museums. The respondents across all three years opined that social tagging helps individuals, beyond artists, to engage with museums and offer their own knowledge and opinions. Reasons to support their views were knowledge and connection. This process sparks conversation and offers additional insights. Many expressed that folksonomy is a way for individuals to connect with one another via social networks; and they can do so in a relaxed manner by sharing their interpretations and opinions of different art works. According to several respondents, users dedicate their time and effort in developing tags; this initiative makes it easy for them to connect and generate conversations with other taggers with similar interests. Another common theme from the responses was emotional expression. Social tagging offers users an opportunity to express how art affects them emotionally. One respondent suggested that it might be possible to build a community of individuals who were emotionally moved by an art collection.

Some differences were noted between the pandemic cohort and the post-pandemic cohorts. As an instance, one respondent belonging to the post-COVID cohort reported: "the sense of ownership and contribution that users experience can increase their engagement with the platform and the content." According to another respondent: "Social tagging/folksonomic terms can assist in connecting anyone with art, artifacts, or exhibits, opening up cultural education and exploration possibilities for anyone who wishes." One respondent stated: "Folksonomies are inherently social, and much of the value that museums can gain from their integration with online collection databases is through increased interaction with, and understanding of, audience." Conversely, some respondents shared their lack of support toward social tagging. According to one respondent: "there can be reasons behind general misinterpretation such as some work is very in-depth that there is a greater meaning behind the art that non-experts cannot formulate." Another respondent shared: "social tagging also undermines the role of museums because it can reduce an object's authenticity since it would focus more on engagement and individualized views."

TABLE 5.1 Potential for Audience Engagement

Response	*Percentage*	*Reasons*
Agreed	100%	1 It inspires audience to use personal imagination to enhance engagement for better ideas. 2 Social tagging is a way that people have learned to create structure in the boundless space of the Internet archival information. 3 This collaborative word bank can help reach people who otherwise would not have connected with certain museum pieces. 4 Rather than being observers, visitors are contributing to the conversation. Social tagging also is educational in how it reveals how different people view and engage with a given piece of art giving museums and curator's insights on how heritage is serving or not serving the community.
Disagreed	0%	N/A

With the opportunity to offer input on artworks and share it online, people are more likely to desire engaging with the museum exhibits. The Generation Z is tech savvy, having grown up with technology. Therefore, it expects to be edutainment (mix of education with entertainment) type of experiences which could be one of the reasons museums are demonstrating keenness to open their exhibits for social tagging (Ann Marshall Personal Communication, March 30, 2020). These social tagging platforms are easily accessible and are user-friendly, thereby making it possible to for anyone to create and post tags online. The views of the sample support recent literature about museum initiatives to evolve and keep up with the changing needs of the society that requires experiential engagement (Chhabra, 2021; Gonsales, 2021; Loureiro et al., 2020; Trant, 2009). Emerging trends indicate that the audiences prefer to communicate and consume information in innovative ways (Chhabra, 2021; Timothy, 2021; West, 2017). Also, the responses indicate their belief that social tagging does not challenge the role of historians; rather, it validates their claims in that the individuals still look up to them for accurate knowledge on a subject. Summary of all responses is presented in Table 5.1.

Museum Abrogation

With regard to the question eliciting views on whether folksonomy is an abrogation of museum responsibility, particularly the pandemic cohort was supportive of this program and did not agree that it compromised the responsibility of the museums. The respondents from this cohort found folksonomy to be advantageous to museums in assisting them to maintain viewer interest through online platforms. When travel restrictions were imposed around the world, those looking to "visit" museums were able to do so online through online exhibitions and other interactive programs offered by the museums. This allowed viewers to post social tags and connect to people via the Internet while offering new ideas and interpretations of different artworks. With the help of folksonomy, museums were able to obtain insights on viewer way of engaging with the exhibit and could use those terms to attract more interest by inviting user-to-user interactions.

Meanwhile, a year later in 2022, museums began to re-open their doors physically to the public. Respondents from the 2022 cohort shared a somewhat different view of how folksonomy was shaping the role of museums (as seen in Table 5.2 and in some notable unedited answers of the respondents). Most pointed out the social tagging "ultimately detracts from the museums' goal." According to some respondents, technology has also resulted in a loss of jobs, particularly right after the pandemic, as the shift into the virtual world continues. Several respondents shared that their jobs were compromised or closed. It was opined that museum employees are trained professionals whose main purpose is to educate others and folksonomy has the ability to take away from that and put their jobs at risk of being renounced.

TABLE 5.2 Possible Abrogation of Museum Responsibility

Response	*Percentage*	*Reasons*
Agreed	24%	1 It can reduce an object's authenticity since it can focus more on engagement and individualized views. 2 There can be reasons behind general misinterpretation such as some work is very in-depth and that there is a greater meaning behind the art that non-experts cannot formulate. 3 While folksonomic terms in museum records sound like a great idea, it is one of the main responsibilities of the museum. Leaving this in the hands of the public would ultimately be leaving too much responsibility that would then be out of the museum's hands.
Disagreed	76%	• Folksonomy allows museum personnel to view the users of art websites and elicit perceptions of what the public may be interested in and in what manner they engage with the work. • Folksonomies are inherently social, and much of the value that museums can gain from their integration with online collection databases is through increased interaction with, and understanding of, audience. • Social tagging/folksonomic terms can assist in connecting anyone with art, artifacts, or exhibits, opening up cultural education and exploration possibilities for anyone who wishes.

Furthermore, those who agreed that folksonomy disrupts the responsibility of museums, also argued that it takes away from the authenticity and reliability of museums. According to one respondent in the 2023 cohort: "If open tagging is permitted, museum records may be subject to vandalism, offensive tags, or tags that are inaccurate representations of the material. It might damage the museum's credibility and reputation." This indicates that as the pandemic became a part of the past and the respondents were distanced from the effects of that catastrophic time, they viewed social tagging as a hindrance in that it took away from the role of museums, which is to deliver accurate and reliable information to visitors.

To further test the concept of social tagging on Generation Z, the respondents were given the social tagging task, as described in the *Method* section. The respondents were asked to find a piece of art or object (that appealed to them) on one of the recommended websites and interpret it, before looking for its actual meaning on the museum website. This exercise directed respondents to develop their own perceptions of an art work and actively engage with the exhibit. It was found that many respondents "misinterpreted" the original meaning of the art piece. However, after reading further into the intention of the artist, they shared that this exercise piqued their interest and they were able to gain a better understanding of the art piece; and they also felt inspired to research more into the original meaning behind the art piece. The respondents shared that they enjoyed reading the actual meaning of the art piece after social tagging the exhibit. In other words, the artwork exercise had facilitated audience engagement and interest as they were inspired to further explore the history and meaning of the piece. Furthermore, many respondents reported that they took great pride in developing their own interpretation of the art because it allowed for the art to be more approachable and there were no limitations on the way they could interpret the art. According to one respondent from the 2021 cohort: "I think that social tagging is useful for audience engagement as it made me think first prior to finding the answer. I was able to appreciate the work more (for longer time), and I was able to walk away with a deeper understanding that" will stay in memory and can be useful "in my own daily life." Rather than being hesitant to think in the shoes of the artist, the interpreters were able to get creative about the exhibit and express their views.

Conclusion

This study is subject to several limitations. It was conducted in a controlled setting and the sample size was small. However, it was able to elicit views over a period of three years, from Generation Z. Therefore, it offers important insights and can be used as a stepping-stone to survey a broader population and across different generations. In the museum context, future studies can look at the manner in which museums use the information generated from the social tags and the commitment of their time and effort to design creative social tagging programs.

In summary, the results of this study illustrate that the majority of Generation Z respondents offer positive feedback on the concept of social tagging and folksonomy. However, some of the respondents (particularly from the post-pandemic cohorts) find social tagging to be unproductive and detrimental for the employees of the conventional museums. These findings depart from the observation of most recent studies which support folksonomy (Gonsales, 2021; Samanta & Rath, 2021; Trant, 2006, 2009). However, for the most part, the respondents express support and the reasons they share correspond with the logic shared in documented literature. It was interesting to note that maximum support for folksonomy was expressed by the respondents during the pandemic year (2021).

Museums have evolved with the advancement in technology to best sustain their purpose and existence and by catering to the needs of their audience. Social tagging is seen, by many, as a form of entertainment that allows for users to gain enrichment. Before the outbreak of COVID-19, when museum exhibitions were not yet fully online, one might have inquired about the popularity of social tagging. The methods of social tagging and folksonomy are likely going to advance over the years and taggers might play an important role to retain audience interest while at the same time foster social bonds with online communities. As mentioned in the literature review, RRC is an emerging style and museums focusing on contemporary art and culture might be interested in capturing real time events and objects as they are being used. Social tagging audience for RRC might be different from the traditional social taggers. Future scholars can examine this line of inquiry.

If used in a strategic manner, folksonomy and social tagging can facilitate civic engagement and social and psychological well-being of the public. Furthermore, by engaging the public in exhibits and artwork showcasing several aspects of sustainability, museums can contribute to contribute to several sustainable development goals such as SDG 3, SDG 16, and SDG 17.

The museums should use participatory approaches grounded in citizen science (CS) and citizen humanities (citizen "science" in the humanities) (Heinisch et al., 2021). According to Pierroux, "CS strives to recognize citizens as stakeholders who may frame issues in quite different ways to the framings used by scientists, policymakers and other experts" (2020, p. 283). In fact, participatory strategies can be used to involve citizens in early initiatives during the collection and documentation phases particularly in regard to the RRC. Pierroux notes that "the democratizing features of knowledge (co-) production, increased dialogue and interactions between scientific institutions and participants, and enhanced citizen engagement and understanding in science and cultural heritage, are acknowledged, with new forms of labor capital in the cultural heritage sector also identified as an emergent and urgent economic and ethical concern for museums and archives" (2020, p. 286). Citizen humanities holds the capability to "combine topical issues in society with academic knowledge, demonstrate the relevance of the humanities for society, and establish a direct link to its members"; and it can also "unlock the potential of embedded, diverse, and culturally sensitive knowledge and play a crucial role in preserving and enriching cultural knowledge" (Heinisch et al., 2021, p. 97). Public participation and engagement can be in the form of "transcribing and annotating text or museum objects, adding (contextual) information from different sources, interpreting digitized documents" (Heinisch et al., 2021, p. 105). A phased approach is required and future studies should explore the potential contributions of this approach.

References

Alves da Veiga, P. (2020). Curating the everywhere museum of everything preprint - universidade aberta. https://repositorioaberto.uab.pt/bitstream/10400.2/12444/3/Curating_the_Everywhere_Museum_of_Everything_PrePrint.pdf

Anderson, S. (2020). Unsettling national narratives and multiplying voices: The art museum as renewed space for social advocacy and decolonization—A Canadian case study. *Museum Management and Curatorship*, *35*(5), 488–531.

Bearman, D., & Trant, J. (2005). Converting scanned images of the print history of the world to knowledge: A reference model and research strategy. *UNESCO between Two Phases of the World Summit on the Information Society*, 423.

Bonney, R., Phillips, T. B., Ballard, H. L., & Enck, J. W. (2016). Can citizen science enhance public understanding of science? *Public Understanding of Science*, *25*(1), 2–16.

Chhabra, D. (2009). Proposing a sustainable marketing framework for heritage tourism. *Journal of Sustainable Tourism*, *17*(3), 303–320.

Chhabra, D. (2021). *Resilience, authenticity and digital heritage tourism*. Routledge.

Chung, J. Y., Chen, C.-C., & Lin, Y.-H. (2016). Cross-strait tourism and generational cohorts. *Journal of Travel Research*, *55*(6), 813–826.

Cosley, D., Baxter, J., Lee, S., Alson, B., Nomura, S., Adams, P., Sarabu, C., & Gay, G. (2009). *A tag in the hand: Supporting semantic, social, and spatial navigation in museums*. Proceedings of the 27th International Conference on Human Factors in Computing Systems, CHI 2009, Boston, MA, USA, April 4–9.

Crompton, J. L., & McKay, S. L. (1997). Motives of visitors attending festival events. *Annals of Tourism Research*, *24*(2), 425–439.

Ding, Y., Jacob, E. K., Zhang, Z., Foo, S., Yan, E., George, N. L., & Guo, L. (2009). Perspectives on social tagging. *Journal of the American Society for Information Science and Technology*, *60*(12), 2388–2401.

Eger, L., Komárková, L., Egerová, D., & Mičík, M. (2021). The effect of COVID-19 on consumer shopping behaviour: Generational cohort perspective. *Journal of Retailing and Consumer Services*, *61*, 102542.

Golder, S. A. & Huberman, B. A. (2006). Usage patterns of collaborative tagging systems. *Journal of Information Science*, *32*(2): 198–208.

Gonsales, F. I. (2021). Social marketing for museums: An introduction to social marketing for the arts and culture sector. *RAUSP Management Journal*, *56*(3), 314–333.

Heinisch, B., Oswald, K., Weißpflug, M., Shuttleworth, S., & Belknap, G. (2021). *Citizen humanities. The science of citizen science*. SpringerLink.

Hekman, E., & Vliet, H. van. (n.d.). *Bringing the past to the present: The use of tagging and storytelling …* Bringing the Past to the Present. The use of tagging and storytelling for the enrichment of digital cultural heritage. https://www.semanticscholar.org/paper/Bringing-the-Past-to-the-Present.-%3A-The-use-of-and-Hekman-Vliet/5129931acb86c052f1c1df089069ebdbc5d3c752

Hudoshnyk, O., & Krupskyi, O. P. (2023). Practices of using rapid response collecting by Ukrainian museums in wartime. *Muzeológia a Kultúrne Dedičstvo*, *11*(2), 5–16.

Hume, M., & Mills, M. (2011). Building the sustainable iMuseum: Is the virtual museum leaving our museums virtually empty? *International Journal of Nonprofit and Voluntary Sector Marketing*, *16*(3), 275–289.

Kabassi, K. (2017). Evaluating websites of museums: State of the art. *Journal of Cultural Heritage*, *24*, 184–196.

Katifori, A., Chrysanthi, A., Vayanou, M., & Antoniou, A.. (2020). Cultural heritage and social experiences in the times of COVID 19.

Komarac, T., Ozretic-Dosen, D., & Skare, V. (2017). Understanding competition and service offer in museum marketing. *Academia Revista Latinoamericana de Administración*, *30*(2), 215–230.

Lewis, G. D. (2019). Museum. *Encyclopædia britannica*, Encyclopædia Britannica. Retrieved from www.britannica.com/topic/museum-cultural-institution

Loureiro, S. M. C. (2020). Virtual reality, augmented reality and tourism experience. *The Routledge handbook of tourism experience management and marketing* (pp. 439–452). Routledge.

Marshall, A. (2020). *What is a museum? A dispute erupts over a new definition*. The New York Times – Online. Retrieved from www.nytimes.com/2020/08/06/arts/what-is-a-museum.html?smid=fbshare&fbclid=IwAR21eZZS_U-8hzrLx7zRxiZy5lCQ8DhP9pjmrbD8ISCsMlmmUUbPp4FyDKg

McKenna, G., & Patsatzi, E. (2005). *SPECTRUM: The UK Museum Documentation Standard*. MDA.

Pierroux, P. (2020). Citizen science, citizen humanities: Relevance for museum research and practice. In P. Hetland, P. Pierroux, & L. Esborg (Eds.), *A history of participation in museums and archives* (pp. 281–296). Routledge.

Samanta, K. S., & Rath, D. S. (2021). Measuring the applicability of user-generated social tags along with expert-generated LCSH descriptors in sociology: A heuristic study. *Annals of Library and Information Studies (ALIS)*, *68*(1), 28–38.

Steve Museum. (2008). An ongoing experiment in social tagging, folksonomy, and museums. https://www.museumsandtheweb.com/biblio/stevemuseum_ongoing_experiment_social_tagging_folkson.html

TEME (n.d.). The museum of everything. Retrieved December 5, 2023, from https://www.musevery.com/#about

Timothy, D. J. (2021). *Cultural heritage and tourism: An introduction* (2nd ed.). Channel View Publications.

Tenenbaum, D. (2020). UU. https://studenttheses.uu.nl/bitstream/handle/20.500.12932/37311/MAAS_Thesis_Daniela%20Tenenbaum.pdf?sequence=1&isAllowed=y

Thangavel, P., Pathak, P., & Chandra, B. (2021). Millennials and Generation Z: A generational cohort analysis of Indian consumers. *Benchmarking: An International Journal*, *28*(7), 2157–2177.

Trant, J. with the participants in the steve.museum project (2006). Exploring the potential for social tagging and folksonomy in art museums: Proof of concept. *New Review of Hypermedia and Multimedia*, *12*(1), 83–105.

Trant, J. (2009). Studying social tagging and folksonomy: A review and framework.

Trant, J., & Wyman, B. (2006, May). Investigating social tagging and folksonomy in art museums with steve.museum. In Collaborative web tagging workshop at WWW2006, Edinburgh, Scotland.

Vander Well, T. (2007). Folksonomy coinage and definition. Retrieved January 2024, from https://vanderwal.net/folksonomy.html

Walsh, J., & Gribbon, D. (1997). (1997). *The J. Paul Getty Museum and its collections: a museum for the new century*. Getty Publications.

West, M. (2017). A review of subject indexing and social tagging projects in art museums. Retrieved on February 12, 2024. University of North Carolina: Carolina Digital Repository.

6

TRADITIONAL-STYLE HOUSES

More Authenticity or Comfort?

Linsi He

Introduction

Suppose you live in a community Xijiang Qianhu Miao village, a special ancient village in the Guizhou province of China, and you decide to operate an inn to accommodate tourists and make money at the same time. Regarding the local traditional wooden architecture, you faced a dilemma: based on the claim that tourists are in pursuit of authenticity (MacCannell, 1973), whether the B&B should be run as authentic as possible or modified so that tourists feel more comfortable, e.g., by installing flush toilets, bathtubs, or putting soundproof construction materials inside traditional wooden walls.

This dilemma seems to have resulted from the inns' special wooden structure. Traditional buildings with wooden structures have various benefits suited to conventional local life, such as being warm in winter and cool in summer, easy to maintain, and moisture-proof in spring (Xiao, 2018). Yet, wooden buildings are not soundproof and difficult to build toilets or bathrooms. Thus, when wooden buildings are transformed into homestays or inns, they might be noisy (Xiao, 2018), especially during tourists season. And if we do not address the question regarding tourists' desire for pursuing authenticity and the potential influence of comfort, we might face similar dilemmas constantly. For example, Nenet dolls, traditionally made yet viewed as "creepy" because of using actual animal parts (Keskitalo et al., 2019, p. 12); or authentic yet uncomfortable places or experiences (Mura, 2015; Wang, 2007; Zhou et al., 2018).

Despite comfort being less studied directly in tourism except in a few (Al-Sakkaf et al., 2021; Al-Sallal et al., 2018; Karimi & Mohammad, 2022; Sheng & Tang, 2015), this potential variable seems to make the previously

DOI: 10.4324/9781003346197-6

mentioned proposition of tourists seeking authenticity less certain. Some research have revealed or mentioned the ambiguity surrounding tourists' demands of authenticity and comfort (Vidon & Rickly, 2018; Zhou et al., 2018). Several explanations are provided; for example, Gnotha and Wang (2015) explained existential authenticity-oriented tourists' empathetic understanding as a reason for their compromise of authenticity seeking. Zhou et al. (2018) nominated these phenomena which are authentic yet unacceptable for tourists as a new type of authenticity, negative authenticity, and claiming that they are unfavorable. Wang (2007) explains tourists' satisfaction with staged authenticity by searching for "home." All these studies provide some insights into understanding this phenomenon.

From the perspective of objective authenticity, this study focuses on exploring whether comfort can be a mediating factor between tourists' desire for organically authentic experience and satisfaction. And in order to explore whether comfort could be a mediating variable, a case study of Xijiang Qianhu Miao village as mentioned above was conducted. It is the largest village inhabited by the ethnic group people, Miao people, around 1,920 households lived there (Wanyan & Luo, 2020) (see Figure 6.1). Their most distinctive building, Diaojiaolou, is a wooden building well suited to the physical geographical environment since ancient times and has become one of the most distinctive tourism resources in China, attracting millions of tourists

FIGURE 6.1 Xijiang Qianhu Miao village

Source: Personal collection of the authors

visiting this place every year. For example, around 8.1426 million made trips there in 2018 (Zhang, 2020). However, as mentioned above, wooden houses are generally not soundproof, and bathrooms are hard to build inside (Xiao, 2018). As both the most spectacular tourism resources and potential accommodations for tourists (many of these buildings are transformed into local B&B), the case study of tourists who visited the Miao village (see Figure 6.1) and stay in the traditional building is well suited to explore tourists' demand for authenticity and the potential influence of comfort.

This study aims at exploring whether tourists' desire for authenticity and satisfaction can be mediated by comfort. Next, we will illustrate related literature, including authenticity and comfort. After that, in the *Methodology* section, content analysis is employed to examine tourists' comment reviews of the traditional-style houses that they have resided in to answer the specific research questions proposed based on the research objective. The result of this study was presented later, followed by a conclusion, discussion, and limitations.

Literature Review

In light of the general research objective above, literature regarding authenticity specifically in the tourism field is first explored. Then, to better understand tourists' desire for comfort, existing studies related to comfort are further explored. As comfort has not been extensively studied in the tourism field, relevant research from other disciplines is also examined to enhance understanding of people's preferences for comfort.

Authenticity in Tourism

Our demand for authenticity has existed for a long time, reflected in its application in a number of disciplines, including marketing, psychology, philosophy, sociology, and anthropology, among which tourism studies is a prominent section ('Andéhn & L'Espoir Decosta, 2020; Bruner, 1994; Cohen, 1988a; Graburn, 1979; Julie & Mitch, 2015; MacCannell, 1973; Newman, 2019; Newman & Dhar, 2014; Peterson, 2013; Reisinger & Steiner, 2006; Rickly, 2022; Vidon & Rickly, 2018; Wang, 1999). Authenticity has long been a popular academic topic in tourism contexts (Boorstin, 1977; Chhabra, 2008, 2019; Chhabra et al., 2003; Cohen, 1979, 1988a, 1988b; Cohen & Cohen, 2012; MacCannell, 1973; Wang, 1999). It is frequently associated with heritage tourism, or tourism in general, and is believed to influence tourists' satisfaction and purchase behaviors (Asplet & Cooper, 2000; Chhabra et al., 2003; Dai et al., 2021; Genc & Gulertekin Genc, 2022). Many tourists enjoy traditional local foods (Sims, 2009), buy traditional crafts (Littrell et al., 1993), overnight in local homestays (Jamaludin et al., 2012), and live a perceived "authentic" local lifestyle. It appears that MacCannell's (1973) assertion was correct; people seek authenticity during travel.

Actually, authenticity is a motivating factor in people's decision to visit a specific destination (Bruner, 1994; Knudsen et al., 2016; Park et al., 2019; Rickly-Boyd, 2012; Vidon & Rickly, 2018). According to MacCannell (1973), tours and pilgrimages are both seeking authentic experiences. The idea of seeing the "real" Spain or Italy, seeing something "unspoiled," or observing whether indigenous people really live the way they are often portrayed is a major touristic topos (Culler, 1981). This will allow tourists to encounter the "native" and become more immersed in the world of otherness. Despite for some researchers, authenticity is just a fantasy that can never be met (Knudsen et al., 2016); it plays a vital role in marketing, as it has been revealed by some studies that tourists' perceived authenticity constitutes an important part of the satisfaction of their cultural experiences, which might further influence their future behavioral intention, for example, revisit intention (Lu et al., 2015) or recommendation (Cevdet Altunel & Erkurt, 2015).

The causal relationship between perception of authenticity and satisfaction has been examined by various researchers (Dai et al., 2021; Domínguez-Quintero et al., 2020; Genc & Gulertekin Genc, 2022; Hede et al., 2014; Lee et al., 2016). Hede et al. (2014) found that perceived authenticity of the museum, the visitor, and the materials affects visitor satisfaction. Lee and Phau (2018) also found that young tourists' object-based authenticity, existential authenticity, and perceived monetary value positively influenced overall perceived value and subsequently, satisfaction. Moreover, instead of treating authenticity as a sole determinant factor, some studies have showed that there are other mediating factors working between tourists' perception of authenticity and their final satisfaction, for example, emotions (Domínguez-Quintero et al., 2019), experience quality (Domínguez-Quintero et al., 2020), and aesthetic experience (Genc & Gulertekin Genc, 2022). Despite these factors, another factor, comfort, seems to be revealed in a few case studies, especially in traditional-style houses.

Comfort in Tourism

Even though comfort is less studied directly in tourism, comfort seems to have been referred to in several case studies, as "amenities" or "familiarity" (Azarya, 2004; Keskitalo et al., 2019; Mura, 2015; Wang, 2007). And various case studies have revealed that the previously mentioned proposition of tourists seeking authenticity appears to be less certain when this potential variable seem to function, as if tourists did not desire as much authenticity as previously imagined. For example, Wang's (2007) study shows that tourists in Lijiang, China, are satisfied with remodeled homestays that abandon certain elements of traditional Naxi architecture in order to accommodate the needs of tourists. Mura (2015) found that tourists in Malaysia were unwilling to trade comfort for less comfortable but more authentic experiences. Tourists have also shown negative attitudes toward authentic traditional culture, which sometimes come with unhygienic environment, poverty, and dilapidation (Zhou et al., 2018). Keskitalo et al. (2019, p. 12) revealed the fact that the doll, which is traditionally made by Nenet girls with a body made from a decorated piece of textile and a head made from a duck beak (to represent a female) or a goose beak (to represent a male), is considered as "creepy" as it used actual animal parts even though it is produced based on traditional heritage and could be seen as "authentic."

Regarding these phenomena, some researchers attributed this phenomenon of seeking less authenticity to tourists' homemaking in tourism or the authenticity sought to be contrary to modern values. Wang (2007, p. 801) suggests that "tourists may not necessarily be seeking the authenticity implicit in UNESCO's definition"; they might also be seeking a sense of home. As a result of attributing tourists' content with customized authenticity to their desire for a "sense of home," it implied that "home" is synonymous with relaxation, comfortable, and convenience, which some researchers might doubt (Cresswell, 2014). As for Zhou et al. (2018), tourists seek less "authenticity" as it might be contrary to the value of modern civilization, which raises the question of how they define modern civilization.

Some other researchers ascribed the phenomenon to a particular group of people. According to Gnotha and Wang (2015, p. 171), existential authenticity-oriented tourists can experience authenticity in the context of the commoditization of culture because they empathize with local people's rights of development and their rights of privacy. Although existential authenticity-oriented tourists claim they can achieve authenticity in a commercialized context, it is unclear whether this conclusion applies to other types of tourists and, more importantly, how to separate them.

In this study, instead of assuming tourists seeking (complete) authenticity or no authenticity, tourists' need for authenticity is understood as a continuum (Chhabra, 2008), and this potential factor, comfort, might affect how authentic tourists seek. Boorstin (1977) had long argued that tourists seek comfort in tourism. In his words, tourists expect more "strangeness and familiarity than the world naturally offers" (Boorstin, 1961/1977, p. 79), an adventure of a lifetime without risks, where "exotic and familiar can be made to order" (Boorstin, 1961/1977, p. 80). Though, the conclusion that tourists are passive, expecting interesting things to happen, and living in a tourist "bubble" far from reality, and seeking no authenticity has gone too far. Azarya (2004) mentioned that even though tourists claim that they wish to immerse themselves in other cultures and places, they are unwilling to give up daily amenities of their familiar surroundings and lifestyles, for example, their hot showers.

The importance of comfort for tourists is also revealed in more recent studies, especially in heritage buildings, districts, or areas (Al-Sakkaf et al., 2021; Al-Sallal et al., 2018; Karimi & Mohammad, 2022; Rutty & Scott, 2010; Sheng & Tang, 2015; Tegar & Suwantoro, 2018; White & White, 2009). As Tegar and Suwantoro (2018, p. 1) said, "People must feel be able to feel comfortable to act rationally in their physical environment ... Without comfort, how good the existing building's architecture cannot be enjoyed." Sheng and Tang (2015) showed that when tourists suffer from excessive traffic noise at certain sties, this lack of acoustic comfort may have a negative impact on promotion of route-based tourism in the long run. The importance of thermal comfort for tourists is more and more researched, and focusing on figuring out the thermal comfort conditions in the tourism sites (Karimi & Mohammad, 2022; Rutty & Scott, 2010). The importance of comfort from social companionship is also mentioned (White & White, 2009). In considering the wide dimensions of this concept, a more thorough and deep understanding of this concept is needed.

Comfort Studies

Research on comfort is important in several disciplines, for example, nursing. The goal of nursing is to provide comfort, thereby relieving pain and distress, and restoring health (Gropper, 1992). Within this context, the meaning of comfort is frequently understood relief from discomfort (or a cause of the state of comfort), a state of ease, and relief from discomfort (Kolcaba, 1991). Furthermore, comfort has also been explored from various other angles, including as a basic human need, a process, or an outcome of nursing (Gropper, 1992; Malinowski & Stamler, 2002). Kolcaba and Kolcaba (1991) further denoted three comfort needs, which are associated with three comfort states: the need for being in a comfortable state and the state sense; the need for relief from discomfort and the relief sense; and the need for personal growth and the renewal sense.

Aside from nursing, ergonomics also heavily relies on it. Increasingly, comfort is seen as a selling point of products, such as chairs, cars, clothing, hand tools, and even airplane tickets (De Looze et al., 2003). Slater (1985, p. 4) defines comfort as "a pleasant state of physiological, psychological and physical harmony between a human being and the environment." According to Slater (1985), the first type of comfort refers to human body's ability to continue functioning, such as breathing. While psychological comfort refers to the ability of the mind to function satisfactorily without our active assistance. According to Slater (1985), physical comfort is easier to grasp since it is more tangible, such as discomfort caused by an itch or wearing the wrong clothing.

As Dumur et al. (2005) point out, comfort can be examined from many scientific perspectives, including psychological, physical, sociological, and technological. Psychologically, comfort refers to a state of quiet enjoyment without worry, disappointment, boredom, financial difficulty, etc. Physical comfort is related to freedom from pain and suffering, relaxation, feeling pleasant and satisfied, and feeling physically well. For example, tourists may be affected by air quality, noise, and crowded environments. Sociologically, comfort is conditioned by belonging to a group, a family, or a culture. Wearing certain clothes can evoke perceptions of being out of place. Lastly, the technological viewpoint focuses on providing pleasure and agreeable sensations. And comfort is provided by technological means, by manufacturing objects that have (or lack) comfortable characteristics, which are measurable.

In tourism context, as revealed previously, comfort is less studied except a few (Al-Sakkaf et al., 2021; Al-Sallal et al., 2018; Chang et al., 2022; Karimi & Mohammad, 2022; Rutty & Scott, 2010; Sheng & Tang, 2015; Tegar & Suwantoro, 2018; White & White, 2009). These studies are generally clustered around psychological aspects (White & White, 2009) and especially physical aspects, for example, thermal comfort (whether thermal conditions in destinations or tourism sites are neither too hot nor too cool) (Al-Sakkaf et al., 2021; Karimi & Mohammad, 2022; Rutty & Scott, 2010), service quality (Otto & Ritchie, 1996), visual comfort (how to ensure visual comfort and avoid discomfort come from direct sunlight or high illuminance levels in heritage buildings) (Al-Sallal et al., 2018), comfort of public facilities (Tegar & Suwantoro, 2018), acoustic comfort (influence of noise on tourism) (Sheng & Tang, 2015), or comfort food (Spence, 2017).

Methodology

Research Sites, Backgrounds, and Specific Research Questions

To explore the relationship between tourists' pursuit of authenticity and their needs for comfort, this study employed the content analysis method to examine TripAdvisor reviews of traditional-style houses in Qianhu Miao Village (see Figure 6.1). Considering the positive image bestowed upon travelers seeking authenticity, and the generally negative tone used for tourists seeking comfort, content analysis of tourists' comment reviews may offer some useful insight.

Traditional wooden-structure buildings, Diaojiaolou, were gradually converted to frame-structure buildings as tourism developed in the first decade of the 21st century. A frame-structure building can provide bathrooms, toilets, and soundproof materials to tourists and residents. However, a transformed building structure may impact the magnificent landscape of traditional wooden-structure buildings. To ensure the "authenticity" of these ethnic inns, the local government issued several regulations concerning their conservation, for example, the Regulation on the Protection Ethnic Cultural

Villages in Miao and Dong autonomous prefecture in southeast Guizhou (黔东南苗族侗族自治州民族文化村寨保护条例) in 2012. As a result of the regulation mentioned above, buildings gradually evolved into a state combined with masonry concrete and wooden structures (Xiao, 2018). Even though local inns used soundproof material, built separate bathrooms, enhanced decorations, and so on, their hybridized structure, still has many "shortcomings" (or, in another way, "authentic" representations of traditional buildings).

Based on objective authenticity, these hybridized local inns differ from being close to the criteria to being completely unrelated to them (it is hard to define the beginning criteria, so, for example, criteria of building houses in the 1980s), leading to different perceptions of authenticities. The more authentic a traditional-style house converted into an inn is, the closer it matches what most people see as authentic Diaojiaolou in the 1980s. The levels of authenticity may vary among people depending on multiple factors. Compared with tourists staying in local hotels, tourists staying in these inns are assumed to be seeking authenticity.

A feeling of comfort is a mix of feelings, perceptions, moods, and situations (Dumur et al., 2005). In light of previous literature, comfort in this study is examined from a physical perspective to answer the research question related to tourists' desire for authenticity and comfort. Unlike comfort, discomfort refers to a tourist's unpleasant state of behavior in response to their physical environment. When tourists comment (positively or negatively) on local inns' physical environment (for example, complaining about soundproofing, or enjoying the amenities), their demand for comfort is revealed. This study does not consider psychological or sociological perspectives on comfort.

Based on the above literature review and the research objective of this study, specific research questions are proposed as follows:

1 Will tourists pursuing authenticity (tourists who stay in local inns and leave reviews online) will have less focus on the physical environment of the inns than on other themes?
2 Will tourists pursuing authenticity be neutral or positive toward inns' physical environment? And for tourists who are dissatisfied with their accommodation experience, will negative attitudes toward the physical environment be the main theme they showed negative attitudes toward?

Content Analysis

A content analysis has been widely employed to explore explicit and covert meaning in texts (Bernard et al., 2017). A text can be any a book, a magazine, an interview transcript, a website, or a social network site (Lawrence, 2011). Concepts, attitudes, beliefs, and so on that are constituted in languages, can also be analyzed using it (Krippendorff, 2018). The line between quantitative and qualitative content analyses is not sharp, according to Schreier (2012). Although qualitative content analysis emphasizes latent meaning, meaning not immediately obvious, quantitative content analysis emphasizes manifest, literal meaning. Compared with other qualitative methods for data analysis that open up data, (qualitative) content analysis reduces the amount of material (Schreier, 2013).

Krippendorff (2018) suggests six significant steps that the analyst needs to proceed: (1) Unitizing: using relevant units definitions; (2) Sampling: using sampling plans; (3) Recording/coding: following coding instructions; (4) Reducing data to manageable representations; (5) Abductively inferring contextual phenomena; (6) Narrating the answers to the research question. The first four components constitute what we may call data collection (or data making, referred to by Krippendorff (2018)). However, they don't have to be arranged in a linear manner. The content analysis design may include iterative loops, or recurrences of components, but it must describe the content details (Krippendorff, 2018).

We selected an online comments sample from local inns on TripAdvisor as a representative sample of all tourists who have stayed at those inns. The reason for choosing TripAdvisor over other online tourism platforms from China is that it is an international tourism platform and may represent perceptions from tourists around the world. With a search for "Leishan county" on TripAdvisor, all accommodations explicitly located in Xijiang Miao Village appeared with comments. Due to the research question limit, hotels or inns with signs of less authenticity, such as fully masonry-concrete buildings with elevators, are excluded from the study. For example, Xijiang Miaohuang Culture Hotel is excluded. Huayu Hotel, despite the name "hotel," is closer to traditional dwellings. Other sources of information are also combined to verify that the comments are all from traditional local inns. By May 19, 2022, comments and reviews from 27 local inns are collected.

Using MAXQDA software, two dimensions are coded based on the research questions: the topics tourists focused on and their attitudes toward them. In reviews, adjectives or conjunctions indicate tourists' attitudes based on their understanding contexts toward these topics. A text that mentions "the beds there are extremely small … utterly unsuitable for a couple …" (by ShomerShabbatFamily) has been coded as "unsatisfied bed," a facility-related code and their attitude has been coded as "negative." Another comment is "Nice view from deck" (by PocahontasRevenge). Regarding topics, the text has been coded as "Ideal view," and regarding attitude, it has been coded as "positive."

Reliability and validity are criteria for evaluating research rooted in quantitative social science methodology (Schreier, 2012). However, neither is a yes-or-no type of criteria. The coding frame is neither reliable nor unreliable, valid or invalid. Instead, it is reliable or valid to a certain degree (Schreier, 2012). To increase the reliability, coefficients of agreements from different coders is one method that is frequently employed to show the intercoder reliability.

A study's coding frame could be regarded as valid if it adequately reflects the concepts in the research questions. Specifically, it involves two contentious areas: the first relates to the validity of coding frames for assessing manifest versus latent content, while the second relates to the scope of the conclusions drawn on the basis of content analysis (Schreier, 2012, p. 175). According to Schreier (2012), it is often justified by making the smallest inferential "leap" when making conclusions about the social situation in which they were created. However, to make inferences from the material to the social situation requires more evidence.

Results

In order to answer the specific research questions, we first presented the general information revealed by all the tourists' comment reviews, namely all the specific aspects (themes) that tourists emphasized. Tourist attitudes toward these aspects (positive, negative, or neutral) are examined later in greater detail. Due to the assumption that tourists who live in local inns pursue authenticity, specific research questions proposed previously could be answered based on the results revealed.

Overview of the Themes Mentioned by Tourists in Online Comment Review

In examining all online comment reviews from 27 local inns, four major themes (with sub-themes comprising 84.68% of the total codes) emerged (see Table 6.1). Tourists who are in pursuit of authenticity focused more on the physical environment of inns or rooms than on other themes based on the four major themes.

Theme 1. Physical Environmental Aspects of Inns or Rooms

The physical environment is the most frequently mentioned theme in tourists' comments (157 codes mentioned by 40 out of 43 comment reviews). It includes their perceptions regarding the benefits or discomforts brought by location, and how the inn is decorated both inside and outside. The most concerned or highlighted themes are "quiet seeking" and, most importantly, the facilities, such as the size of rooms or bathrooms. During the coding process, there are a few situations where tourists mentioned how they avoided

TABLE 6.1 Major Themes Tourists Focused based on Comment Reviews

Major themes	*Sub-themes*	*Number of sub-themes codes*	*Percentage of sub-themes codes*	*Number of comment reviews from tourists (number of documents)*	*Percentage of tourists who mentioned the theme*
Physical environment aspects of inns or rooms	Facilities*	89	27.64%	33	76.74%
	Location*	38	11.80%	25	58.14%
	Quiet seeking*	18	5.59%	16	37.21%
	Decoration*	14	4.35%	12	27.91%
Services from staff or owner of inns	Evaluation of services*	28	8.70%	19	44.19%
	Various services provided*	18	5.59%	10	23.26%
Activity or performances	Enjoy ideal view	32	9.94%	24	55.81%
	Festival/performance or other activities*	6	1.86%	5	11.63%
	Enjoy good rural environment	1	0.31%	1	2.33%
Food and products	Evaluation of the taste of the meals*	16	4.97%	12	27.91%
	Choices of meals/ products/environment of having meals*	11	3.42%	9	20.93%
	Prices of meals or products*	2	0.62%	2	4.65%
General feelings	Worthwhile	6	1.86%	6	13.95%
	Disappointed	5	1.55%	5	11.63%
	Comfortable	5	1.55%	5	11.63%
	Not worthwhile	3	0.93%	3	6.98%
Price	Relatively reasonable/ inexpensive	5	1.55%	5	11.63%
	Relatively expensive	2	0.62%	2	4.65%
	Overpriced	2	0.62%	2	4.65%
	Discount negotiation	1	0.31%	1	2.33%
Recommendations or not	Recommend	6	1.86%	6	13.95%
	Not recommend	3	0.93%	3	6.98%
	Worth stay but not long	1	0.31%	1	2.33%
Factors influencing choices decision of inns	Influenced by photos and comment reviews from others	5	1.55%	5	11.63%
	Not much choice	1	0.31%	1	2.33%
Perception regarding authenticity/ commodification	Modern/commodified*	2	0.62%	2	4.65%
	Authentic*	2	0.62%	2	4.65%
Total		322	100.00%		

Note: * suggest that this sub-theme has further sub-themes.

those discomforts, for example, by selecting certain rooms to avoid noise. Even though this phrase did not reveal the tourists' discomfort, it did indicate tourists' concern about the importance of quiet environments.

> The shared bathroom need[needs] some renovation. It has a big hole if im [I'm] not mistaken it is a hole for a window, it is covered by a piece of fabric.
>
> *(TH6[1])*

> The hotel is impossible to find without a guide - maze-like alleys. With the luggage it is difficult to get up the narrow streets and the uneven steps, so only carry-on luggage.
>
> *(TH19)*

> The hotel is decorated with Miao features,
>
> *(TH18)*

> However, because it is a wooden building, the sound insulation effect is relatively poor, and you can hear louder voices upstairs and downstairs.
>
> *(TH24)*

> I was woken up the next day, and several tourists had been talking outside. I didn't wake up from the thunder yesterday, and waves of tourists in the morning waked me up.
>
> *(TH9)*

> … but it's quiet and nice because it doesn't face the street.
>
> *(TH26)*

Theme 2. Services from Staff or Owners of Inns

The services from staff or inn owners are the second major theme that tourists mentioned in their comment reviews (46 codes mentioned by 26 comment reviews out of 43). It includes two sub-themes. One is various specific services they praised or hoped for, such as picking up services, introductions or guide services, and other services. The other sub-theme is tourists' evaluation of these services, such as the quality of the services and the attitude of staff (e.g., whether they were friendly).

> Also, every time the female owner of the inn met, she smiled and greeted me, which made me very happy.
>
> *(TH38)*

> The child of the staff (daughter like a family business?) Brings me from the first floor immediately while I'm out of breath, and the service feels warm.
>
> *(TH25)*

Theme 3. Activity or Performance

Activity or performance is the third theme that emerged from tourists' comment reviews (39 codes were mentioned by 24 comment reviews out of 43). Most of the tourists emphasized enjoying the splendid view there, and a few also mentioned other activities, such as joining the tourism performance, joining the local festivals, participating in the local night market, cooking classes, and so on.

> Hotel is on the hillside with a good view of Visasvis [vis-as-vis]to the old town.
>
> *(TH19)*

> Also the temporary market along the river (not the established shops) is a great place to buy silver Miao jewellery.
>
> *(TH30)*

Theme 4. Food and Products

The final major topic emerged from the tourist comment reviews (29 codes mentioned by 16 comment reviews out of 43). The codes are most associated with the choice of meals or products or even the environment. The other principal subcategory is related to their evaluation of these meals.

> For a westerner, it is a terrible breakfast …
>
> *(TH29)*

> No coffee, no tea, soup and soy milk in a whole …
>
> *(TH28)*

> Instead, you are allowed to bring your own food and drinks.
>
> *(TH5)*

Other Themes

Theme 5. General Feelings

Around 17 comment reviews (19 codes) represented their feelings toward this accommodation experience. Some tourists express their feelings at certain moments, for example, feeling comfortable, or feeling disappointed. Some other tourists expressed their emotions based on certain comparisons, thus making judgments such as whether they were worthwhile or not worthwhile.

> The hotel was very comfortable,
>
> *(Th30)*

> We were rather disappointed, however, in part because we had picked the place on the basis of the pictures on booking.com, which are very misleading: they vastly exaggerate the size of the rooms and the beds, for one thing.
>
> *(TH13)*

> But when I arrived at the 5th floor and saw the view from the observatory on each floor, The tiredness is blown away!
>
> *(TH25)*

> In short, there may be better for the price.
>
> *(TH7)*

Theme 6. Recommendations or Not

Ten tourists (10 codes) also expressed their recommendations or not recommending suggestions.

> I will not go back there. Not recommend to those who are sound sensible.
>
> *(TH10)*

> I will continue to visit next time and recommend it when friends come. Xijiang Miao Village is very beautiful, xxx is definitely the most beautiful hotel in Xijiang Miao Village, and it is highly recommended.
>
> *(TH22)*

> Overall, it's worth to stay for a nite but not for[a] long vacation.
>
> *(TH6)*

Theme 7. Price

Ten comment reviews (10 codes) are related to price.

> The price was reasonable
>
> *(TH25)*

> Overpriced for the quality. The rooms here are on the more expensive side relative to other hotels in Xijiang area, even compared to others nearby that have good hilltop views.
>
> *(TH1)*

Theme 8. Factors Influence Tourists' Purchase Decision

There are six comment reviews (6 codes) that mentioned factors that have influenced their purchase decision.

> We booked this hotel based on the reviews we saw on trip.com as there aren't many reviews for Xijiang Miao Village that we wanted to go to.
>
> *(Th16)*

Theme 9. Tourists' Perception of Authenticity

There are four tourists mentioned their perceptions regarding authenticity in the comment review (4 codes).

> Xijiang is a large Miao village - restored and perhaps a little contrived rather than authentic - but beautiful nevertheless.
>
> *(TH30)*

> You can see the 'real' village life with dogs, cats and chickens in the streets and neighbors building their wooden houses.
>
> *(TH5)*

Tourists' Attitudes toward Various Topics

The relationship between tourists' pursuit of authenticity and potential influence of comfort is further examined by coding co-occurrence between various topics tourists choose and their attitudes toward those topics. And by coding co-occurrences, the codes which occur more often or less often could be evaluated based on the size of the circles. Figure 6.2 shows how many document segments are attached to any two codes.

Positive Attitudes toward Various Themes

According to tourist comments, local views (32 codes from 24 comment reviews), inn locations (23 codes from 19 comment reviews), and services (including evaluation and provision of services) are most generally positive. Facilities have some positive codes (around 25.84%), but most are negative

	Neutral	Negative	Positive	SUM
Topic				0
General feelings		•	•	19
Recommendation or not	•	•	•	10
Physical environment of inns or r				0
Facilities	•	•	•	89
Location	•	•	•	38
Quiet seeking	•	•	•	18
Decoration	•		•	14
Services from staff or owner of i				0
Service evaluation		•	•	28
Various services provided	•	•	•	18
Activity/performance/				0
Enjoy ideal View			•	32
Festival/ performance or othe			•	6
Enjoy good rural environment			•	1
Food and products				0
Meals taste evaluation	•	•	•	16
Price of meals or products			•	2
Choice of meals/products/en	•	•	•	11
Price	•	•	•	10
Factors influence the B&B choice	•	•	•	6
Perception regarding authenticity		•	•	4
Σ SUM	63	85	174	322

FIGURE 6.2 Code co-occurrences table

Code co-occurrences table: It shows the frequency of two codes (one code is from the Topic, such as *physical environment*. And the other code is from the Attitude, such as *positive*) which occur at the same quotation.

(around 41.57%) or neutral (32.58%). Comments regarding local views are all positive. For example,

> Wooden hostel with [the] best viewpoint of the Xijiang Miao village.
>
> *(TH3)*

> The view from our quadròom is magnificent.
>
> *(TH6)*

When it comes to sub-topics of homestay location, tourists who hold a neutral attitude tend to state the difficulty of accessing the inns with (heavy) luggage due to uneven ground or a lack of elevators. Tourists who hold a positive attitude tend to emphasize the convenience brought by the location of the inns, which are close to various other tour destinations or ideal for enjoying the landscape.

> The straight-line distance between the inn and the viewing deck is really only 10 meters. I climbed to the top of the mountain with my bag on my back and without taking a tourist bus, I was too tired.
>
> *(TH4)*

> It is conveniently located along the river in the center of the village, close to theaters and rice terrace spots, and in an area with many restaurants.
>
> *(TH26)*

> From the room, I could see the mountain opposite the square and the houses lined up at the foot of the mountain.
>
> *(TH42)*

In positive comments regarding facilities, cleanliness is frequently mentioned (5 codes). Other aspects are also mentioned, such as Wi-Fi (4 codes).

> The room is cottage-like and clean
>
> *(TH26)*

> Along the way, I feel that this [B&B]is the cleanest
>
> *(TH12)*

> Wifi was OK for surf[ing] the internet.
>
> *(TH3)*

Within the comments regarding sub-topics of decoration, almost all the tourists showed a positive attitude. Tourists expressed a supportive attitude regarding (the creative) decoration of B&B with local Miao characteristics.

> The drawing on the wall [is] quite attractive and creative, good job!
>
> *(TH6)*

> Yes, it's not only tiny but also characteristic.
>
> *(TH10)*

For tourists who showed a positive attitude regarding services, as reviewed previously. Thus, it will not be further mentioned here.

Negative Attitude toward Various Themes

Of all the negative comments (85 codes) left by tourists, 55 codes are related to the physical environment of inns, including various facilities deficiencies or wooden structures not being soundproof (it has been revealed before, thus not further discussed here). For the sub-theme of facilities (37 codes out of 55), tourists showed negative attitudes toward various aspects, including bathrooms (14 codes). Some tourists mentioned that they do not have private bathrooms, and some said its small size.

> Cons: 1.No private bathroom.
>
> *(TH6)*

> ... and restroom is quite cramped.
>
> *(TH1)*

> ... and the bathroom/shower/squat toilet combo was about as wide and long as my arms stretched out.
>
> *(TH2)*

After room size, negative comments are also related to the simple facilities within the bathrooms (6 codes).

> When you shower [,] half the bathroom is wet[,] and you have nothing to put your toiletry bag on.
>
> *(TH29)*

> The bathroom (freezing) has a shower placed above the toilet seat (see photo). On the door[,] we ask to place the towels outside when we use it [them]. So we shower straddling the toilets and we go out to get the towels afterward[s].
>
> *(TH28)*

Other aspects generated from tourists' comments include unsatisfied rooms (9 codes) due to small, dark, or damp rooms, small balconies, or unsatisfied facilities (3 codes) such as small beds or low-quality housewares.

> Room is relatively small ...
>
> *(TH1)*

> It was small with very low ceilings.
>
> *(TH2)*

> Watch out for the "twin" rooms: the beds there are extremely small[,] and utterly unsuitable for a couple.
>
> *(TH13)*

> The twin bed was hard[,] ...
>
> *(TH2)*

The above results suggest that tourists who desire authenticity are still very much concerned about the physical environment of the inns or the rooms in the inns, especially the facilities. And tourists showed negative attitudes toward physical environments that they may not find comfortable, for example, small bathrooms, simple facilities, and in-soundproofing, even though these might constitute manifestations of "authenticity" based on the criteria of traditional dwelling in the 1980s.

Considering the situation that tourists who in pursuit of authenticity might be willing to overcome the difficulties brought by "authenticity," meaning even though tourists showed negative attitudes toward the physical environment, they still rated this experience as satisfying. In order to further explore tourists' attitudes toward the importance of the physical environment in their pursuit of authenticity, in this study, the final rating of their experience in these inns as additional data is used to evaluate whether they are satisfied with their experience (3 tourists do not rate their experience are excluded in this section). Comments with ratings less than five are considered that their experiences are not entirely satisfying.

These comments are generated from 31 comments from tourists who rated their accommodation experience less than 5 (see Figure 6.3). Similar to the previous results, negative comments mostly focused on the physical environment of homestays or rooms (49 out of 79 codes). In other words, tourists with low ratings for their accommodation experience were more likely to have negative attitudes toward the physical environment. This result suggests that even though tourists are seeking authenticity, the physical environment, specifically the discomfort of facilities, might influence their satisfaction with the accommodation experience. Thus, there is a high chance that comfort might function as a mediating variable between tourists' desire of authenticity and their final satisfaction.

Conclusions and Discussions

This study examines the relationship between tourists seeking authenticity and the potential influence of comfort, specifically, whether comfort can serve as a mediating variable between tourists' desire for authentic experiences and their ultimate satisfaction. To investigate this relationship, content analysis is employed to answer the following three questions: Whether tourists who

Code System	Neutral	Negative	Positive	SUM
Topic				0
General feelings		•	•	14
Recommendation or not	•	•	•	5
Physical environment of inns or r				0
Facilities	●	●	•	66
Location	•	•	●	29
Quiet seeking	•	•	•	16
Decoration	•		•	9
Services from staff or owner of i				0
Service evaluation		•	•	17
Various services provided	•	•	•	6
Activity/performance/				0
Enjoy ideal View			●	23
Festival/ performance or othe			•	3
Enjoy good rural environment			•	1
Food and products				0
Meals taste evaluation	•	•	•	12
Price of meals or products				0
Choice of meals/products/env	•	•	•	9
Price	•	•	•	8
Factors influence the B&B choice	•	•		5
Perception regarding authenticity		•	•	2
Σ SUM	47	79	99	225

FIGURE 6.3 Themes generated from comment reviews by tourists

Tourists who rated their accommodation experience less than 5: showing frequency of two codes which occur at the same quotation from comments posted by tourists who rated their accommodation experience less than 5.

are in pursuit of authenticity will pay less attention to the physical environment of inns than to other themes in the reviews? And will tourists, who are in pursuit of authenticity, have a positive or at least a neutral attitude toward the inns' physical environment? Will negative attitude toward the physical environment constitute the main theme among those who are dissatisfied with their accommodation experience?

According to this study, tourists who are seeking authenticity were highly concerned with the physical environment, including facilities of inns and rooms, suggesting that tourists who are seeking authenticity are still concerned about physical comfort, especially comfort of facilities and the acoustic comfort. This finding corroborated Gropper's (1992) work that comfort is a basic human need pursued by all human beings, including tourists who desire authenticity in this study. As Tegar and Suwantoro (2018) point out, comfort is indispensable for creating a friendly neighborhood or city. Without it, tourists will not be able to enjoy the beauty of a building. According to Tegar and Suwantoro (2018, p. 2), comfort can be provided in a variety of ways, including orientation, color, traffic calming, vegetation, bins, signage, public toilets, seating areas, information centers, parking spaces, pedestrian paths, and public transportation. For traditional-style houses, facilities like room size, bed quality, and housewares play an important role in tourists' comfort in the physical environment.

Also, this study found that tourists seeking authenticity may show negative attitudes toward uncomfortable physical environments, despite the fact that these uncomfortable environments may be part of what is referred to as "authenticity" based on certain criteria, for example, the poor soundproofing due to wooden structure. Therefore, tourists' quest for authenticity does not (totally) counteract discomfort's influence. Their final satisfaction could be affected by it as well. Among tourists with low ratings for their accommodation experience, negative attitudes toward the physical environment constitute the majority of themes in the review. It suggests that perceived physical comfort might influence tourists' accommodation satisfaction, which might further influence their following-up behaviors, for example, their overall ratings of their accommodations, the length of time they intend to stay, or whether they intend to purchase specific authentic products.

For example, despite the beautiful view, helpful and friendly staff, and creative decoration of the homestay, the negative attitude brought by the discomfort of the facilities made the tourist, TH6, to comment that "*Overall it's worth to stay for a night but not for long vacation.*" Similar finding has also been revealed in Mura's (2015, p. 231) study, "'authentic-triggering' experiences should not last long periods of time as guests seem to be keen to comprise their comforts only for short periods of time." And the homestay providers are suggested to identify the "perfect" balance between authentic/uncomfortable and inauthentic/comfortable experiences. Facing discomfort caused by undesirable climatic conditions, tourists might either alter the timing of their trip or avoid the destination altogether (Rutty & Scott, 2010).

Finally, based on the above results and discussions, this study argues that instead of placing authenticity seeking and comfort as the opposite ends, emphasizing one as the manifestation of encroachment of the other, comfort can be used as a mediating factor between tourists' desire for authentic experience and their final satisfaction. Thus, for different tourists who are in pursuit of a certain level of authenticity, their final satisfaction with the visit may attribute a significant section to their comfort of the experience in a heritage accommodation. This proposes another interesting question for future studies, to what extent will visitors forgo the quest for an authentic experience beyond their comfort zone.

Knowing that tourists are influenced by comfort while seeking authenticity is key to providing authenticity-related or comfort-related interventions. A local practitioner can make their tourism offerings more popular by maintaining comfort levels and increasing "authenticity" levels or increasing comfort level under the current authenticity level. When comfort level remains the same, creatively blending more local and traditional elements into architecture, souvenirs, crafts and food or certifying souvenirs and crafts as hand-made or made by indigenous people can help (Skinner et al., 2020). Similarly, under present authenticity, service providers can maximize comfort by designing food that is more palatable, providing more food choices, making homestays more comfortable, or increasing the functional value of the tourism product.

In addition, acknowledging the influence of comfort for tourists who desire authenticity could relieve the stress and pressure imposed by "authenticity" on local people in destinations, specifically for the local inn operators in this study. Within the context of globalization, differences, "primitiveness," marginality, or even poverty in certain areas, especially in remote, developing areas, have often been used to represent the difference with other "standardized" areas to attract visitors to the destination (Azarya, 2004). In these areas, tourism is frequently considered the best option for economic progress (Cole, 2007a; Salazar, 2009). However, with the gradual development of the local destination, and more access to modernization (such as electricity or modern bathrooms), the destination is no longer considered "primitive" or "authentic" (compared with the stereotyped representation of local destination). This access to modernization in these places is gradually losing its appeal, which used to attract tourists (Cole, 2007b). This conflict brought about by socio-economic integration with cultural distinctions remaining authentic is considered to be a paradox or a dilemma for people in developing areas (Cole, 2007b, 2007a; Salazar, 2009; Wanyan & Luo, 2020).

Thus, measures taken to improve the welfare of local people or tourists may not necessarily discourage tourists from visiting this destination. After all, as Cohen (1988a) points out, tourists' criteria for making comparisons might gradually evolve over time and what they perceive as authentic may change as well. Just as Novelli and Tisch-Rottensteiner (2012) revealed, tour operators and local tourism destinations could inform tourists about destinations' progress before they travel to. Also, it may also not appeal to tourists if it adheres to certain criteria so desperate level that it sacrifices the welfare of locals. After all, based on the above discussion, tourists themselves are seeking a balance between comfort and authenticity.

Limitations and Recommendations for Future Research

This chapter contributes to explore whether comfort can serve as mediating factor between their desire for authenticity and satisfaction. The influence of comfort, which manifested as concern for the physical environments in traditional-style houses, including facilities, services, food, and acoustic comfort. This study is not without limitations. First, these comment reviews could only represent a particular sub-group of tourists, namely those who would like to post comments on the websites. Thus the generalization of conclusions might lead to a reliability problem (Volo, 2010), so additional data sources should be provided for further analysis. Researchers could do follow-up studies with tourists who tended to leave comments and those who did not leave a comment review. Besides, this study made two assumptions. The first assumption is that assuming traveler ratings and comments could well represent travelers' attitudes or perceptions. Five is considered the most satisfying; one is considered the least satisfying. The second assumption is that tourists who stay in homestays or local inns are considered as tourists who are in pursuit of authenticity. There is, however, a chance that tourists' ratings do not represent what they considered, or that tourists who choose to stay in homestays or inns may be attributed to various other factors. For example, they do not have money. So, follow-up studies to track these tourists' perceptions would enhance this study's validity.

Note

1 Tourists whose comment reviews are examined in this study are labeled as TH1, TH2, ..., TH43 (using the first two letters of the chapter title traditional-style houses).

References

Al-Sakkaf, A., Mohammed Abdelkader, E., Mahmoud, S., & Bagchi, A. (2021). Studying energy performance and thermal comfort conditions in heritage buildings: A case study of Murabba Palace. *Sustainability*, *13*(21), Article 21. https://doi.org/10.3390/su132112250

Al-Sallal, K. A., AbouElhamd, A. R., & Dalmouk, M. B. (2018). UAE heritage buildings converted into museums: Evaluation of daylighting effectiveness and potential risks on artifacts and visual comfort. *Energy and Buildings*, *176*, 333–359. https://doi.org/10.1016/j.enbuild.2018.06.067

Andéhn, M., & L'Espoir Decosta, J. N. P. (2020). Authenticity and product geography in the making of the agritourism destination. *Journal of Travel Research*, *60*(6), 1282–1300. https://doi.org/10.1177/0047287520940796

Asplet, M., & Cooper, M. (2000). Cultural designs in New Zealand souvenir clothing: The question of authenticity. *Tourism Management*, *21*(3), 307–312. https://doi.org/10.1016/S0261-5177(99)00061-8

Azarya, V. (2004). Globalization and international tourism in developing countries: Marginality as a commercial commodity. *Current Sociology*, *52*(6), 949–967.

Bernard, H. R., Wutich, A., & Ryan, G. W. (2017). *Analyzing qualitative data: Systematic approaches*. SAGE Publications.

Boorstin, D. J. (1961/1977). *The image: A guide to pseudo-events in America*. Atheneum.

Bruner, E. M. (1994). Abraham Lincoln as authentic reproduction: A critique of postmodernism. *American Anthropologist*, *96*(2), 397–415.

Cevdet Altunel, M., & Erkurt, B. (2015). Cultural tourism in Istanbul: The mediation effect of tourist experience and satisfaction on the relationship between involvement and recommendation intention. *Journal of Destination Marketing & Management*, 4(4), 213–221. https://doi.org/10.1016/j.jdmm.2015.06.003

Chang, J., Morrison, A. M., Lin, C.-T., & Lin, S. H.-H. (2022). Exploring the emotions and well-being of food neophobic travelers in the consumption of comfort food. *Food Quality and Preference*, *96*, 104443. https://doi.org/10.1016/j.foodqual.2021.104443

Chhabra, D. (2008). Positioning museums on an authenticity continuum. *Annals of Tourism Research*, *35*(2), 427–447.

Chhabra, D. (2019). Authenticity and the authentication of heritage: Dialogical perceptiveness. *Journal of Heritage Tourism*, *14*(5–6), 389–395.

Chhabra, D., Healy, R., & Sills, E. (2003). Staged authenticity and heritage tourism. *Annals of Tourism Research*, *30*(3), 702–719.

Cohen, E. (1979). Rethinking the sociology of tourism. *Annals of Tourism Research*, *6*(1), 18–35.

Cohen, E. (1988a). Authenticity and commoditization in tourism. *Annals of Tourism Research*, *15*(3), 371–386. https://doi.org/10.1111/j.1745-6592.2010.01279.x

Cohen, E. (1988b). From tribal costume to pop fashion: The "Boutiquisation" of the textiles of the hill tribes of Northern Thailand. *Studies in Popular Culture*, *11*(2), 49–59.

Cohen, E., & Cohen, S. A. (2012). Authentication: Hot and cool. *Annals of Tourism Research*, *39*(3), 1295–1314.

Cole, S. (2007a). Beyond authenticity and commodification. *Annals of Tourism Research*, *34*(4), 943–960.

Cole, S. (2007b). *Tourism, culture and development: Hopes, dreams and realities in East Indonesia*. Channel View Publications.

Cresswell, T. (2014). *Place: An introduction*. John Wiley & Sons.

Culler, J. (1981). Semiotics of tourism. *The American Journal of Semiotics*, *1*(1/2), 127–140.

Dai, T., Zheng, X., & Yan, J. (2021). Contradictory or aligned? The nexus between authenticity in heritage conservation and heritage tourism, and its impact on satisfaction. *Habitat International*, *107*, 102307. https://doi.org/10.1016/j.habitatint.2020.102307

De Looze, M. P., Kuijt-evers, L. F., & Van Dieën, J. (2003). Sitting comfort and discomfort and the relationships with objective measures. *Ergonomics*, *46*(10), 985–997. https://doi.org/10.1080/0014013031000121977

Domínguez-Quintero, A. M., González-Rodríguez, M. R., & Paddison, B. (2020). The mediating role of experience quality on authenticity and satisfaction in the context of cultural-heritage tourism. *Current Issues in Tourism*, *23*(2), 248–260. https://doi.org/10.1080/13683500.2018.1502261

Domínguez-Quintero, A. M., González-Rodríguez, M. R., & Roldán, J. L. (2019). The role of authenticity, experience quality, emotions, and satisfaction in a cultural heritage destination. *Journal of Heritage Tourism*, *14*(5–6), 491–505.

Dumur, E., Barnard, Y., & Boy, G. (2005). Designing for comfort. In D. de Waard, K.A. Brookhuis, and C.M. Weikert (Eds.), *Human factors in design safety and management* (pp. 111–127). Shaker Publishing B.V.

Genc, V., & Gulertekin Genc, S. (2022). The effect of perceived authenticity in cultural heritage sites on tourist satisfaction: The moderating role of aesthetic experience. *Journal of Hospitality and Tourism Insights*, *6*(2), 530–548. https://doi.org/10.1108/JHTI-08-2021-0218

Gnotha, J., & Wang, N. (2015). Authentic knowledge and empathy in tourism. *Annals of Tourism Research*, *50*, 170–172.

Graburn, N. H. H. (1979). *Ethnic and tourist arts: Cultural expressions from the fourth world*. University of California Press.

Gropper, E. I. (1992). Promoting health by promoting comfort. *Nursing Forum, 27*(2), 5–8. https://doi.org/10.1111/j.1744-6198.1992.tb00905.x

Hede, A.-M., Garma, R., Josiassen, A., & Thyne, M. (2014). Perceived authenticity of the visitor experience in museums: Conceptualization and initial empirical findings. *European Journal of Marketing, 48*(7/8), 1395–1412. https://doi.org/10.1108/EJM-12-2011-0771

Jamaludin, M., Othman, N., & Awang, A. R. (2012). Community based homestay programme: A personal experience. *Procedia - Social and Behavioral Sciences, 42*, 451–459. https://doi.org/10.1016/j.sbspro.2012.04.210

Julie, M., & Mitch, G. (2015). How aspects of a wine's place affect consumers' authenticity perceptions and purchase intentions: The role of country of origin and technical terroir. *International Journal of Wine Business Research, 27*(1), 61–78. https://doi.org/10.1108/IJWBR-01-2014-0002

Karimi, A., & Mohammad, P. (2022). Effect of outdoor thermal comfort condition on visit of tourists in historical urban plazas of Sevilla and Madrid. *Environmental Science and Pollution Research, 29*(40), 60641–60661. https://doi.org/10.1007/s11356-022-20058-8

Keskitalo, E. C. H., Schilar, H., Heldt Cassel, S., & Pashkevich, A. (2019). Deconstructing the indigenous in tourism. The production of indigeneity in tourism-oriented labelling and handicraft/souvenir development in Northern Europe. *Current Issues in Tourism, 24*(1), 16–32. https://doi.org/10.1080/13683500.2019.1696285

Knudsen, D. C., Rickly, J. M., & Vidon, E. S. (2016). The fantasy of authenticity: Touring with Lacan. *Annals of Tourism Research, 58*, 33–45.

Kolcaba, K. Y. (1991). A taxonomic structure for the concept comfort. *Image: The Journal of Nursing Scholarship, 23*(4), 237–240. https://doi.org/10.1111/j.1547-5069.1991.tb00678.x

Kolcaba, K. Y., & Kolcaba, R. J. (1991). An analysis of the concept of comfort. *Journal of Advanced Nursing, 16*(11), 1301–1310. https://doi.org/10.1111/j.1365-2648.1991.tb01558.x

Krippendorff, K. (2018). *Content analysis: An introduction to its methodology*. Sage Publications.

Lawrence, N. W. (2011). *Basics of social research: Qualitative and quantitative approaches*. Pearson.

Lee, S., & Phau, I. (2018). Young tourists' perceptions of authenticity, perceived value and satisfaction: The case of Little India, Singapore. *Young Consumers, 19*(1), 70–86. https://doi.org/10.1108/YC-07-2017-00714

Lee, S., Phau, I., Hughes, M., Li, Y. F., & Quintal, V. (2016). Heritage tourism in Singapore Chinatown: A perceived value approach to authenticity and satisfaction. *Journal of Travel & Tourism Marketing, 33*(7), 981–998.

Littrell, M. A., Anderson, L. F., & Brown, P. J. (1993). What makes a craft souvenir authentic? *Annals of Tourism Research, 20*(1), 197–215.

Lu, L., Chi, C. G., & Liu, Y. (2015). Authenticity, involvement, and image: Evaluating tourist experiences at historic districts. *Tourism Management, 50*, 85–96. https://doi.org/10.1016/j.tourman.2015.01.026

MacCannell, D. (1973). Staged authenticity: Arrangements of social space in tourist settings. *American Journal of Sociology, 79*(3), 589–603.

Malinowski, A., & Stamler, L. L. (2002). Comfort: Exploration of the concept in nursing. *Journal of Advanced Nursing, 39*(6), 599–606. https://doi.org/10.1046/j.1365-2648.2002.02329.x

Mura, P. (2015). Perceptions of authenticity in a Malaysian homestay—A narrative analysis. *Tourism Management, 51*, 225–233. https://doi.org/10.1016/j.tourman.2015.05.023

Newman, G. E. (2019). The psychology of authenticity. *Review of General Psychology*, *23*(1), 8–18.

Newman, G. E., & Dhar, R. (2014). Authenticity is contagious: Brand essence and the original source of production. *Journal of Marketing Research*, *51*(3), 371–386.

Novelli, M., & Tisch-Rottensteiner, A. (2012). Authenticity versus development: Tourism to the hill tribes of Thailand. In O. Moufakkir, & P. M. Burns (Eds.), *Controversies in tourism* (pp. 54–72). CABI Wallingford.

Otto, J. E., & Ritchie, J. R. B. (1996). The service experience in tourism. *Tourism Management*, *17*(3), 165–174. https://doi.org/10.1016/0261-5177(96)00003-9

Park, E., Choi, B.-K., & Lee, T. J. (2019). The role and dimensions of authenticity in heritage tourism. *Tourism Management*, *74*, 99–109.

Peterson, R. A. (2013). *Creating country music: Fabricating authenticity*. University of Chicago Press.

Reisinger, Y., & Steiner, C. J. (2006). Reconceptualizing object authenticity. *Annals of Tourism Research*, *33*(1), 65–86.

Rickly, J. M. (2022). A review of authenticity research in tourism: Launching the annals of tourism research curated collection on authenticity. *Annals of Tourism Research*, *92*, 103349.

Rickly-Boyd, J. M. (2012). 'Through the magic of authentic reproduction': Tourists' perceptions of authenticity in a pioneer village. *Journal of Heritage Tourism*, *7*(2), 127–144.

Rutty, M., & Scott, D. (2010). Will the Mediterranean become "too hot" for tourism? A reassessment. *Tourism and Hospitality Planning & Development*, *7*(3), 267–281. https://doi.org/10.1080/1479053X.2010.502386

Salazar, N. B. (2009). Imaged or imagined? Cultural representations and the "touristification" of peoples and places. *Cahiers d'études Africaines*, *49*(193–194), 49–71.

Schreier, M. (2012). *Qualitative content analysis in practice*. Sage Publications.

Schreier, M. (2013). Qualitative content analysis. In U. Flick (Ed.), *The SAGE handbook of qualitative data analysis* (pp. 170–183). SAGE Publications Ltd.

Sheng, N., & Tang, U. W. (2015). Spatial techniques to visualize acoustic comfort along cultural and heritage routes for a world heritage city. *Sustainability*, *7*(8), Article 8. https://doi.org/10.3390/su70810264

Sims, R. (2009). Food, place and authenticity: Local food and the sustainable tourism experience. *Journal of Sustainable Tourism*, *17*(3), 321–336. https://doi.org/10.1080/09669580802359293

Skinner, H., Chatzopoulou, E., & Gorton, M. (2020). Perceptions of localness and authenticity regarding restaurant choice in tourism settings. *Journal of Travel & Tourism Marketing*, *37*(2), 155–168. https://doi.org/10.1080/10548408.2020.1722785

Slater, K. (1985). *Human comfort*. Charles C Thomas Publisher.

Spence, C. (2017). Comfort food: A review. *International Journal of Gastronomy and Food Science*, *9*, 105–109. https://doi.org/10.1016/j.ijgfs.2017.07.001

Tegar, G., & Suwantoro, N. (2018). How to improve the comfort of Kesawan Heritage Corridor, Medan City. *IOP Conference Series: Earth and Environmental Science*, *126*(1), 012204. https://doi.org/10.1088/1755-1315/126/1/012204

Vidon, E. S., & Rickly, J. M. (2018). Alienation and anxiety in tourism motivation. *Annals of Tourism Research*, *69*, 65–75.

Volo, S. (2010). Bloggers' reported tourist experiences: Their utility as a tourism data source and their effect on prospective tourists. *Journal of Vacation Marketing*, *16*(4), 297–311.

Wang, N. (1999). Rethinking authenticity in tourism experience. *Annals of Tourism Research*, *26*(2), 349–370.

Wang, Y. (2007). Customized authenticity begins at home. *Annals of Tourism Research*, *34*(3), 789–804.
Wanyan, W., & Luo, Y. (2020). *The old house is "not suitable for living" and only "good to see", who will come to see if there is no old house?* Xinhua Daily Telegraph. http://www.xinhuanet.com/politics/2020-10/19/c_1126627418.htm
White, N. R., & White, P. B. (2009). The comfort of strangers: Tourists in the Australian outback. *International Journal of Tourism Research*, *11*(2), 143–153. https://doi.org/10.1002/jtr.723
Xiao, Y. (2018). Change of architectural landscape within the perspective of evolution: The case of Miao Village in Guizhou. *Guangzhou University (Social Science Edition)*, *17*(4), 75–83.
Zhang, T. (2020). *Catch the train of tourism development, Xijiang Miao villages get rich and enter the fast lane (搭上旅游发展列车 西江苗寨致富步入快车道)*. https://cn.chinadaily.com.cn/a/202012/22/WS5fe2d33da3101e7ce9736f01.html
Zhou, Q., Zhang, J., Zhang, H., & Li, X. R. (2018). Is all authenticity accepted by tourists and residents? The concept, dimensions and formation mechanism of negative authenticity. *Tourism Management*, *67*, 59–70.

7

TRIBAL TOURISM IN INDIA

A Regulatory-based Approach towards Empowerment of Tribes

Rijuta Sawhney, Alka Maheshwari, and S.R. Gnanasambandan

Introduction

Tribal communities in India have long been custodians of unique cultural heritage and traditional knowledge, living in harmony with nature and fostering a distinct way of life. However, the advent of modernization and the growth of tourism industry have brought both opportunities and challenges for these indigenous groups. Tribal tourism, as a niche segment, promises to provide travellers with authentic cultural experiences while potentially contributing to the socio-economic empowerment of tribal populations. Yet, concerns about the exploitation of indigenous communities and the preservation of their cultural identity have sparked the need for a regulatory-based approach that ensures sustainable development and empowerment in the context of tribal tourism. One such framework that has gained prominence in this context is the World Indigenous Tourism Alliance's (WINTA) model.

This research delves into the application of the WINTA framework as a case study in the tribal region of Kinnaur, India, focusing on the needs and aspirations of the local tribe, the Kinnauris.

The Kinnauris, an indigenous tribe residing in the picturesque Kinnaur district of Himachal Pradesh, India, have long preserved their unique cultural heritage, deeply rooted in their pristine surroundings. As tourism emerged as a potential avenue for economic growth, the Kinnauri community faced the dual challenge of harnessing the benefits of tourism while safeguarding their cultural integrity and environmental balance. In response to such concerns, the WINTA framework can be introduced as a guiding principle for tribal tourism development in the region.

DOI: 10.4324/9781003346197-7

At the heart of the WINTA model lays a holistic approach to sustainable tribal tourism development, a blueprint that has proven instrumental in empowering the Kinnauri tribe and preserving their cultural heritage. The first critical component of the model is community involvement and decision-making. By engaging the Kinnauri tribe in the formulation of tourism policies and activities, the WINTA framework acknowledges the invaluable insights and perspectives of the local community. This participatory approach empowers the Kinnauris, enabling them to shape tourism offerings that align with their cultural values and aspirations, and ensures that tourism becomes a means of preserving and celebrating their traditional way of life.

Closely linked to community involvement is the emphasis on culturally sensitive tourism practices. The WINTA model encourages the integration of indigenous knowledge, customs, and practices into tourism experiences. This authenticity not only attracts travellers' seeking genuine interactions but also allows the Kinnauri community to share their rich cultural heritage with the world on their own terms. Thus, tourism becomes a platform for mutual learning and appreciation, promoting cultural exchange and understanding.

A fundamental pillar of the WINTA framework is the implementation of revenue-sharing mechanisms. By ensuring that the economic benefits derived from tourism activities are equitably distributed among the Kinnauri community, the model fosters a sense of ownership and responsibility towards sustainable tourism practices. These financial gains can be reinvested in improving local livelihoods, supporting education and healthcare initiatives, and preserving traditional crafts and skills, further empowering the tribe to chart their path towards progress.

Moreover, the WINTA model underscores the significance of capacity-building programmes. By providing training and skill development opportunities, the framework enables the Kinnauri tribe to actively participate in the management and operation of tourism ventures. This capacity-building process reduces reliance on external actors and empowers the community to take charge of their tourism initiatives, enhancing the sustainability of these ventures in the long run.

Research Questions:

- How effectively does the application of the WINTA framework, in conjunction with governmental regulations, address the socio-economic empowerment of the tribes in the context of tribal tourism development?
- What role do governmental regulations play in mitigating potential exploitation of indigenous communities and preserving their cultural identity within the framework of tribal tourism?

- To what extent does the collaboration between governmental bodies and indigenous communities facilitate sustainable tribal tourism practices, as exemplified in the Kinnaur region?
- How can governmental regulations be optimized to balance the promotion of tribal tourism for economic growth with the imperative of safeguarding the cultural integrity and environmental balance of indigenous communities?

Literature Review

The literature review section of this research provides a comprehensive examination of tribal tourism in India, focusing on the regulatory frameworks, benefits, challenges, and the role of governmental authorities in its development. Through an exploration of academic works, reports, and policy documents, this review aims to elucidate the transformative potential of tribal tourism while highlighting the significance of regulatory mechanisms in safeguarding the interests of indigenous communities. By delving into the major themes and concerns surrounding tribal tourism within the Indian context, this review sets the stage for further analysis and discussion on the subject.

Tribal Tourism

Tribal tourism in India is a multifaceted domain that has garnered increasing attention in recent years. It represents an evolving approach towards empowering India's diverse tribal communities through the medium of tourism while simultaneously addressing regulatory aspects. India is home to a plethora of indigenous tribal groups, each possessing unique cultures, traditions, and ecological knowledge. The engagement of these communities in the tourism sector offers the potential for economic and social empowerment, but it also necessitates a careful regulatory framework to safeguard their interests.

This literature review delves into the multifarious dimensions of tribal tourism in India, focusing on the regulatory framework that governs its operations. Through an exploration of academic works, reports, and policy documents, this review seeks to elucidate the transformative potential of tribal tourism while underscoring the significance of regulatory mechanisms in ensuring the welfare and preservation of these marginalized communities. By examining existing scholarship, it aims to provide a comprehensive understanding of the complexities and nuances associated with tribal tourism in India.

The central aim of this literature review is to delineate the major themes, issues, and concerns surrounding tribal tourism within the Indian context. This entails an examination of the regulatory initiatives designed to manage and guide the development of tribal tourism, considering aspects like cultural preservation, economic development, environmental conservation, and social empowerment. Moreover, this review will also explore the challenges and controversies that have emerged in the process of integrating tribal communities into the tourism sector and how regulatory frameworks are evolving in response to these issues.

As the tourism industry continues to grow in India, it is imperative to ensure that tribal tourism does not merely serve as a commercial venture but as a vehicle for tribal development and cultural preservation. Therefore, this literature review contributes to the ongoing discourse by offering an extensive analysis of the existing literature, highlighting gaps in knowledge, and offering insights into the regulatory measures that can drive positive change in the realm of tribal tourism. Through this exploration, the review aims to provide a foundation for further research and policy development that can harness the potential of tribal tourism while safeguarding the rights and dignity of India's tribal communities.

Tribal Tourism within India

India is one of the largest tribal populations' globally and second largest tribal population afterwards the continent of Africa. Tribal tourism is important in many respects at both ideological and empirical levels. Tribal tourism can create opportunities for learning tribal traditions and cultures (Telang & Bokde, 2019). The focus of tribal tourism is to strengthen the capacity of tribal communities and individual entrepreneurs to develop such enterprises and to participate in the global tourism market. One of the key principles of tribal tourism is that the local population is placed at the centre of tourism development interventions in terms of participation in the decision-making process one benefit sharing (Suresh, 2021).

As per this study, the importance of tribal tourism, as an instrument of economic development and employment generation, particularly in remote and backwards, is to be given due recognition (Kumar, 2016). "It is evident that global demand for indigenous tourism experiences is growing at a faster pace, it is high time that Indian tourism is benefitted from the opportunity" (Sibi & Anjaneya Swamy, 2015).

In the year 2014–15, a scheme known as Swadesh Darshan Scheme was launched by Ministry of Tourism towards the development of theme-based Circuits such as religious tourism, heritage, and also tribal tourism. Particularly, tribal circuits are to be developed as a path leading towards development and promotion of tribal areas (Sengupta, 2018).

Benefits of Tribal Tourism to the Tribes

Tribal tourism has enormous benefits which can be brought to the tribes through a mapped out plan such as:

Economic Benefits – Tourism has a huge potential to bring in economic benefits to an area if planned and developed. It can be a key driver of employment, directly through jobs generated by tourism such as in the hotels, tourist guides, souvenir stores, tourist information centres, homestays and indirectly through jobs created indirectly such as property managers, manufacturing for these tourism products and creation of services. From employment generation comes income generation to the local people. Also, multiplier effect gradually takes place in the economy. Thus, due to the rapid globalization it is almost cardinal that even the tribes are self-sustaining.

Socio-cultural benefits – Tourism opens up the way for the people to get acquainted with each other which fosters understanding and peace among the communities. As mutual understanding develops a sense of empathy arises as well which leads to a reduction in prejudices. Following, economic benefits flow in there is a gradual development in the infrastructure or in the blossoming of the particular area under which the facilities can be used by the local people as well. Tourism also raises awareness of the worth and significance of cultural, natural and archaeological sites. It instils pride in one's local and national heritage.

Environmental benefits – Cognizance towards protecting the environment may arise to preserve it from the negative impacts. Foreign tourists can sometimes provide helpful ideas for reducing environmental stress and promoting sustainable tourism. The money may be utilized to expand the area's resources and improve its natural beautification. Studies may be put into action along with research activities that may study the detrimental effects of tourism on the environment through which strategies and aid for ecological/environmental conservation can be adapted.

Challenges Being Faced by Tribes

However, there are certain challenges being faced by the Indian tribes which can be further hampered by the development of tourism activities. These include the following (Figure 7.1).

Displacement: The tribal people are forcefully being evicted from their lands due to certain reasons such as liberalization, privatization, and globalization. Due to this their lands are being occupied for development projects.

- Resource exploitation: The tribal people in earlier had an extortionate control over the resources which were generally in forests such as water, land, forest resources and so on. However, now these resources are subject to exploitation to meet the demands of the huge population.

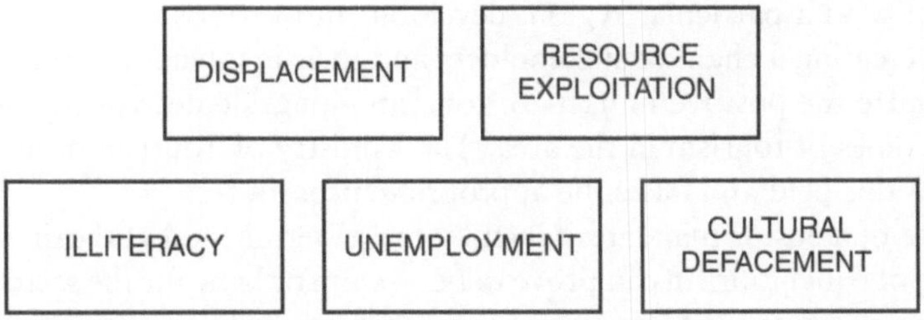

FIGURE 7.1 Challenges being faced by tribes

- Illiteracy: As per the 2011 census survey 59% of the tribal people were found to be literate. Literacy is essential since it helps them be a part of the integrated development process while knowing their rights and awareness of their surroundings.
- Unemployment: The percentage of tribal people working in the modern enterprises in negligible. Over 40% of tribal people are working in the over exploitative sector industries. Unemployment further leads to consequences such as poverty.
- Cultural Defacement: Due to open access to various tribal areas without any policy, no control over the interaction taking place, the tribes are losing their own cultural values and traditions. An example of this is the Jarawa Tribe whose population is steadily declining.

These are some of the broad issues that Indian tribes confront, but they are also at the heart of the challenges that tribal tourism development encounters. The issues include a lack of participation by tribal members and a lack of representation from them.

Although, it must be noted that such tourist activities can become a catalyst for other challenges involving their culture becoming a commodity for the tourists, which can be related to Cultural defacement. Tribal people are concerned that their cultural traditions, morals, and festivals may become a source of entertainment tailored to visitors' requirements. Vices such as exploitation, objectification and culture shock poses a challenge to the development of tribal tourism as it occurs in an uncontrolled manner.

Role of Government Authorities in Development of Tribal Tourism

(Srivastava, 2021) quotes that, "carefully formulated tribal tourism can act as a powerful medium towards social and economic inclusion of Tribals." Additionally, various studies stress that the governmental authorities play a vital role when it comes upon the development of tribal tourism. The government formulates policies, makes plans related to infrastructure development, accessibility to a particular region, development of tourism in a sustainable manner, creating a chain of stakeholders and their involvement and ensuring to maximize the positive impacts of tourism alongside decreasing the negative outcomes of tourism in the area. The Ministry of Tourism in India is in charge of this field and takes the appropriate procedures.

Failure of steps or mandates taken by the governmental authorities in the planning of tribal tourism can prove to be a boom or bane for the sector. It can be ascertained that the tribal regions are an immensely sensitive area which requires major steps to be taken that ensures no harm is brought to the tribes (Chang et al., 2018). In 2014–15, the government launched the Swadesh Darshan scheme which is a central sector scheme instigated by the Ministry of Tourism for the amalgamated development of theme-based Circuits in India.

In yet another case study of Gond Tribes, which is a major Tribe in India, certain suggestions have been proposed towards the development through the actions which can be implemented by the government such as establishment of an action committee, with participation from state government and concerned public and private departments to integrate tribal tourism with the mainstream planning for tourism. A market analysis could be done to connect the requirements of tourists and the expectations of the Gond Tribes. Projects can be started through collaborations between the public and commercial sectors and ideal areas for launching them can be identified or be allocated by the government. A thorough feasibility study must be undertaken in order to examine all of the benefits and drawbacks without detracting from the primary goal of tribal community development. Lastly, Monitoring and assessment should be done formally and on a frequent basis so that negative aspects are focused upon. As a result, we may conclude that the government plays a significant role in the development of tribal tourism and are a significant stakeholder in their development.

Four Hs of Tribal Tourism Development

Various authors have elaborated upon the four Hs of tribal tourism in regard to their development. These four Hs, which encompass Heritage, Hospitality, Handicrafts, and Harmony (Smith, 1996), can serve as a comprehensive approach to promote tribal tourism in India (Pratt et al., 2013). Here is how it has an important role in fostering empowerment and sustainable development for tribal communities.

Heritage

Heritage is the foundation of tribal tourism, as it revolves around preserving and celebrating the unique cultural and historical aspects of indigenous communities. Indigenous heritage includes traditional practices, rituals, art forms, languages, and sacred sites (Goeldner & Gartner, 1995). The preservation and promotion of this heritage play a vital role in the empowerment of tribes by not only retaining their identity but also generating income through cultural tourism. Initiatives such as tribal museums, heritage trails, and cultural festivals have been successful in attracting tourists to tribal areas while also educating them about the rich history and traditions of these communities (Smith, 1996).

Hospitality

Hospitality in tribal tourism is an essential element that can contribute to the sustainable development of tribal communities (Notzke, 2004). Indigenous communities are often known for their warm and inclusive hospitality, which can create unique and authentic experiences for tourists (Arora & Sharma, 2022). Community-based homestays, guided tours by tribal members, and cultural exchanges facilitate interaction between tourists and tribes. This fosters cultural understanding and economic opportunities for tribal members, including income generation from providing accommodation, guiding services, and selling local products (Notzke, 2004).

Handicrafts

Handicrafts are a significant source of income for tribal communities, and they play a pivotal role in tribal tourism development. Traditional craftsmanship, including weaving, pottery, jewellery-making, and textile production, allows tribal artisans to showcase their skills and cultural identity (Dyer et al., 2003). These handicrafts can be marketed to tourists as authentic souvenirs, and the revenue generated can uplift the economic status of tribal communities. The promotion of these products can also help in preserving traditional knowledge and skills that might otherwise be at risk of fading away (Srivastava, 2021).

Harmony

Harmony, in the context of tribal tourism, refers to the coexistence and collaboration between tourists and tribal communities (Butler & Pearce, 2003). It is essential to ensure that tourism activities are sustainable and respectful of indigenous cultures and environments. The promotion of responsible and ecotourism a practice helps maintain the balance between economic development and environmental conservation (Chen, 2019). Tribes must be active participants in decision-making processes to protect their rights and interests. The government, non-governmental organizations, and tribal cooperatives often play a critical role in facilitating this harmony (Verma & Murdia, 2017).

Informed Consent and Will of Tribes

In the context of tribal tourism development in India, the principles of informed consent and respecting the will of tribal communities are pivotal to ensure that initiatives are both ethically and sustainably carried out (Maraña et al., 2023). The concept of informed consent extends to several critical stages of tribal tourism projects. Firstly, it is essential during the project's initiation, ensuring that the tribal community in question is made fully aware of the proposed plans, potential impacts, and implications—both positive and negative (Rao, 2016). Obtaining the community's consent at this stage is foundational in recognizing their autonomy and right to make informed decisions about their involvement in tourism-related activities (Caplan & Moreno, 2011).

Furthermore, informed consent is vital when tribal communities participate directly in tourism initiatives, such as offering hospitality services, guiding tours, or marketing their handicrafts. Clear agreements should be established, outlining roles, responsibilities, compensation, and safeguards to protect the interests and rights of tribal participants (Fredericks, 2016).

In the context of tribal tourism development in India, the principles of informed consent and respecting the will of tribal communities are pivotal to ensure that initiatives are both ethically and sustainably carried out (Johnston, 2014). The concept of informed consent extends to several critical stages of tribal tourism projects. Firstly, it is essential during the project's initiation, ensuring that the tribal community in question is made fully aware of the proposed plans, potential impacts, and implications—both positive and negative. Obtaining the community's consent at this stage is foundational in recognizing their autonomy and right to make informed decisions about their involvement in tourism-related activities (McGregor & Tsosie, 2021).

Furthermore, informed consent is vital when tribal communities participate directly in tourism initiatives, such as offering hospitality services, guiding tours, or marketing their handicrafts (Whyte et al., 2011). Clear agreements should be established, outlining roles, responsibilities, compensation, and safeguards to protect the interests and rights of tribal participants (Fitch et al., 2022).

Beyond the concept of informed consent, respecting the will of tribal communities is equally important. This involves recognizing their right to self-determination and self-governance. It empowers tribal communities to make decisions regarding the extent of their participation in tourism activities, as well as under what conditions (Barelli, 2012). This approach respects the unique cultural heritage and traditions of each tribe and acknowledges that they should determine how and to what degree they wish to share their culture with tourists (Chu, 2015). Additionally, respecting tribal will entails ensuring that the benefits derived from tourism activities are directed in accordance with the priorities and aspirations of the community itself, be it investments in education, healthcare, infrastructure, or other endeavours deemed essential by the community (Caplan & Moreno, 2011).

Benefits and Challenges Associated with Tribal Tourism

This topic has been extensively studied across diverse settings globally, revealing a complex tapestry of benefits and challenges. At its core, tribal tourism serves as a conduit for profound cultural exchange, offering visitors immersive experiences into the rich tapestry of indigenous traditions, languages, rituals, and art forms (Smith, 1996). For indigenous communities, this presents an invaluable opportunity to share their heritage with the world, fostering a sense of pride and preserving their cultural identity in the face of globalization (Buckley, 2012). Economically, tourism can serve as a vital lifeline for these communities, providing much-needed income through the sale of handicrafts, guided tours, and cultural performances, thus bolstering local economies and enhancing livelihoods (Dyer et al., 2003). Moreover, the exposure garnered from tourism can amplify advocacy efforts for indigenous rights and environmental conservation on both local and international platforms, potentially influencing policy decisions and garnering support for community-led initiatives (Verma & Murdia, 2017).

However, alongside these benefits, tribal tourism also poses a myriad of challenges, often rooted in the delicate balance between cultural preservation and commercialization. There exists a palpable risk of cultural commodification, where indigenous traditions and artefacts are commercialized for mass consumption, potentially diluting their authenticity and reducing them to mere commodities for tourist consumption (Chang et al., 2021). This process can lead to the distortion or misrepresentation of indigenous cultures, perpetuating stereotypes and eroding the integrity of traditional practices. Furthermore, the influx of tourists into tribal areas can exacerbate existing socio-economic disparities within communities, as unequal distribution of tourism benefits, exploitation by intermediaries, and overreliance on tourism revenue can deepen existing inequalities and foster dependency (Chouhan, 2022).

Environmental concerns also loom large in the realm of tribal tourism, with increased foot traffic often placing strains on fragile ecosystems and natural resources. Unsustainable tourism practices, such as habitat degradation, waste generation, and disruption of wildlife habitats, can have far-reaching consequences on both the environment and the traditional ways of life of indigenous communities intimately connected to their land (Ishii, 2012). Additionally, prolonged exposure to tourism can inadvertently contribute to the erosion of indigenous cultures, as external influences and pressures to conform to tourist expectations may lead to the gradual loss of traditional practices, languages, and knowledge systems among younger generations (Musavengane, 2019).

Navigating these intricate dynamics necessitates a multifaceted approach that prioritizes the rights, needs, and aspirations of indigenous communities while promoting sustainable and responsible tourism practices (Doğan & Timothy, 2019). This entails fostering genuine partnerships and collaboration between indigenous stakeholders, government agencies, and tourism operators to ensure that tourism initiatives are culturally sensitive, economically equitable, and environmentally sustainable (Huang et al., 2015). By fostering a deeper understanding and appreciation for indigenous cultures and fostering mutually beneficial relationships between tourists and host communities, tribal tourism has the potential to become a powerful force for positive change, fostering greater cross-cultural understanding, respect, and solidarity in an increasingly interconnected world (Bott, 2018).

Literature Gap

Despite the growing body of research on tribal tourism in India, several substantial gaps persist within the existing literature, emphasizing the need for further investigation and exploration. One significant gap lies in the assessment of long-term impacts and sustainability. While many studies shed light on the immediate effects of tribal tourism, such as economic benefits and cultural preservation, there is a dearth of research examining how these impacts evolve over time. Understanding the long-term sustainability of such tourism ventures is vital, including whether economic benefits persist and whether there are unintended consequences such as cultural commodification or the gradual erosion of cultural authenticity.

Another crucial gap pertains to the study of ethical and responsible tourism practices in tribal areas. While there is a growing acknowledgement of the importance of responsible tourism, there is a distinct lack of comprehensive research examining how ethical principles are practically integrated into tribal tourism projects. This gap suggests the need to investigate the actual implementation and effectiveness of responsible tourism practices in safeguarding the rights and autonomy of tribal communities, as well as their role in promoting community empowerment and sustainable development.

Furthermore, the literature has yet to delve deeply into the effectiveness of regulatory frameworks and policies governing tribal tourism. While these regulations are often mentioned, there is a shortage of research on their real-world impact. It is crucial to understand whether existing policies effectively protect the rights of tribal communities, mitigate potential negative consequences, and contribute to cultural and economic development.

A pressing gap in the literature relates to the extent of community participation and decision-making in tribal tourism. While the concept of community-based tourism is discussed, the degree to which tribal communities are actively involved in shaping and controlling the tourism activities on their land remains underexplored. Investigating the dynamics of community participation and its impact on tribal empowerment is essential for advancing our understanding of this field

Research Methodology

The research philosophy adopted for this study was interpretivism. Interpretivism focuses on understanding the subjective meanings and interpretations of individuals and emphasizes the social and cultural context in which the research is conducted. In the context of tribal tourism in India, interpretivism allowed to gain insights into the perspectives and experiences of the local population, specifically the Kinnaura tribe residing in Kinnaur region. As the primary data collection instrument used in this study was a structured questionnaire. The questionnaire was designed based on extensive literature review. It included both closed-ended and open-ended questions to gather quantitative and qualitative data. The questionnaire covered aspects such as the effectiveness of regulatory measures, the impact on the empowerment of tribes, and the overall experience of tribal tourism.

For sampling, purposive sampling was put into effect to specifically target the locals from the Kinnaur region belonging to the Kinnaura tribe. The purposive sampling technique allowed selecting participants who possess the specific knowledge and experience related to tribal tourism and regulatory policies in the region. The sample size for this study consisted of 100 questionnaires collected from the locals.

The data collection procedure involved visiting the Kinnaur region and approaching the locals belonging to the Kinnaura tribe. The researchers explained the purpose of the study and sought their voluntary participation. The questionnaires were administered to the participants, and they were given sufficient time to complete the survey. To ensure that the sample is representative of the population, several measures were taken. Firstly, there was focus on selecting participants from the Kinnaura tribe, which is the target population for this study. Secondly, the researchers employed purposive sampling to select individuals with relevant knowledge and experience in tribal tourism and regulatory policies. This helped ensure that the sample consisted of participants who were familiar with the subject matter. Additionally, efforts were made to include participants from various age groups, genders, and socio-economic backgrounds, to capture a diverse range of perspectives within the Kinnaura tribe.

Thereafter, collected data were analysed using the SPSS software. Descriptive statistics, such as frequency distributions, mean scores, and standard deviations, were computed to summarize the quantitative data. The qualitative data from the open-ended questions were analysed thematically to identify common themes and patterns in participants' responses. The analysis aimed to provide a comprehensive understanding of the participants' perceptions and experiences related to tribal tourism and the regulatory measures.

By employing these research methodologies, the study aimed to gather accurate and reliable data from the locals belonging to the Kinnaura tribe in Kinnaur. The interpretive approach, along with the use of SPSS for data analysis, allowed for a comprehensive examination of the regulatory-based approach towards the empowerment of tribes in tribal tourism in India.

Results

Based on the objectives of the research paper, the data has been gathered from 100 respondents consisting of tribal people involved in trial tourism-related activities in Kinnaur. The respondents' profile is presented in Table 7.1.

Principal Component Analysis (PCA) was employed for extracting the data to determine the factors used to evaluate the perception of tribals towards tourism development in social and cultural context. A five-point Likert scale was used based on the following scale 5=Strongly Agree, 4=Agree, 3=Neutral, 2=Disagree, and 1=Strongly Disagree to evaluate each variable. The Barlett's test of Sphericity was significant (0.00 $p<0.05$). The Kaiser-Meyer-Olkin (KMO) was found to 0.934 which indicates that the sample is good enough for PCA. The PCA with Varimax rotation of the twelve variables resulted in a three-component solution that explains 58.478% of the total variance. Only the factors with Eigen values greater than 1were accepted.

The Components Matrix is the output of the Exploratory Factor Analysis process that lists the loadings of each of the scale items on each of the three components. Valid components having scale item loadings of 0.5 and above (Hair et al., 2007) and scale items with the highest loading on that

TABLE 7.1 Respondents' Demographic Profile

Age	*Frequency*	*Percentage*	*Gender*	*Frequency*	*Percentage*
Below 20	21	21	Male	48	48
21–30	51	51	Female	52	52
31–40	6	6	**Occupation**	**Frequency**	**Percentage**
41–50	15	15	Student	57	57
Above 50	7	7	Employed	27	27
			Unemployed	16	16

TABLE 7.2 Descriptive Statistics

Factor 1: Promotion of Tribal Tourism	
Tribal tourism can lead to Community Empowerment	.676
Adequate Policy Frameworks leads to tourism promotion	.619
Inter-relationship between tribes and the tourists exists	.592
Showcasing handicrafts	.632
Factor 2: Traditional Aspect of Tribal Society	
Tribal Tourism helps in Cultural Preservation	.418
Tribal Tourism leads to Cultural Exchange	.623
Heritage of the tribal society is conserved	.679
Authentic local food enriches the experience.	.444
Factor 3: Sustainable Development in Tribal Area	
Infrastructure and Accessibility	.576
Accommodation and hospitality service	.369
Amenities and Activities	.549
Strategies for developing tribal tourism in the aspects of economy, society and environment in a sustainable way.	.791

component (Wixom & Todd, 2005). This Components Matrix is subsequently rotated using Varimax rotation to assist interpretation of its results (Malhotra, 2007), displaying only loadings of 0.5. Large commonalities indicate that a large number of variances have been accounted for by the factor solution. Varimax rotated factor analytic results for factor perception of tribals towards tourism development. The three factors shown in Table 7.3 have been derived from Rotation.

The standards indicate that data is suitable for factor analysis. Each factor needs to be assigned a name or label to characterize it and aid its interpretation (Tabachnick & Fidell, 2007). Each of the factors that have been extracted via PCA in the EFA process of this research data is displayed in Table 7.2.

Four variables (Community Empowerment, Policy Framework, Interrelation between tribal and tourism, Handicrafts) regarding promotion of tribal tourism were loaded in the first component and accounted for 24.583 of the total variance. The first factor was labelled as Promotion of tribal tourism (Table 7.3).

The second factor labelled Traditional aspect of tribal society loaded with Cultural Preservation, Cultural Exchange, Heritage Conservation, Authentic local food explained 22.127 of the total variance. The third factor interpreted as Sustainable Development in Tribal Area consist of four variables explained 11.769 of the variances. This factor involved (Infrastructure & Accessibility, Accommodation & Hospitality, Activities and Amenities, Sustainable Development of economy, society and environment)

TABLE 7.3 EFA Results

Factor[a]	*Initial Eigenvalues*[b]			*Extraction Sums of Squared Loadings*[f]			*Rotation Sums of Squared Loadings*[g]		
	Total[c]	*% of variance*[d]	*Cumulative %*[e]	*Total*	*% of variance*	*Cumulative %*	*Total*	*% of variance*	*Cumulative %*
1	6.249	52.076	52.076	5.851	48.759	48.759	2.950	24.583	24.583
2	1.229	10.246	62.322	.806	6.719	55.478	2.655	22.127	46.710
3	.719	5.992	68.313	.360	3.000	58.478	1.412	11.769	58.478
4	.613	5.109	73.423						
5	.561	4.676	78.099						
6	.503	4.192	82.291						
7	.471	3.927	86.218						
8	.389	3.240	89.458						
9	.368	3.066	92.524						
10	.328	2.735	95.259						
11	.317	2.645	97.904						
12	.252	2.096	100.000						

[a] The initial number of factors is the same as the number of variables used in the factor analysis. However, not all 12 factors will be retained, only the first three factors will be retained.

[b] Eigenvalues are the variances of the factors. Because we conducted our factor analysis on the correlation matrix, the variables are standardized, which means that the each variable has a variance of 1, and the total variance is equal to the number of variables used in the analysis, in this case, 12.

[c] This column consists of the eigenvalues. The first factor will always have the highest value and account for the most variance , and the next factor will account for as much of the left over variance as it can, and so on. Hence, each successive factor will account for less and less variance.

[d] This column contains the percent of total variance accounted for by each factor.

[e] This column contains the cumulative percentage of variance accounted for by the current and all preceding factors. For example, the third row shows a value of 68.313. This means that the first three factors together account for 68.313% of the total variance.

[f] The number of rows in this panel of the table correspond to the number of factors retained. In this example, we requested that three factors be retained, so there are three rows, one for each retained factor. The values in this panel of the table are calculated in the same way as the values in the left panel, except that here the values are based on the common variance. The values in this panel of the table will always be lower than the values in the left panel of the table, because they are based on the common variance, which is always smaller than the total variance.

[g] The values in this panel of the table represent the distribution of the variance after the varimax rotation. Varimax rotation tries to maximize the variance of each of the factors, so the total amount of variance accounted for is redistributed over the three extracted factors.

Discussions and Analysis

In the discussion section, we delve deeper into the implications of our study's findings within the broader context of tribal tourism development, drawing upon existing literature to enrich our analysis and provide a comprehensive understanding of the complex dynamics at play. By examining the perceptions of tribal communities towards tourism development in the Kinnaur region, our research offers valuable insights into the multifaceted dimensions of tribal empowerment, cultural preservation, and sustainable development within the realm of tribal tourism.

Through the application of PCA on responses from 100 tribal individuals engaged in tribal tourism activities, we uncovered three distinct factors that encapsulate tribal perceptions regarding different facets of tourism development. These factors not only shed light on the attitudes and priorities of tribal communities but also resonate with broader themes and trends documented in the literature on tribal tourism.

The first factor, labelled "Promotion of Tribal Tourism," underscores the critical role of community empowerment, supportive policy frameworks, meaningful interactions between tribes and tourists, and the promotion of indigenous handicrafts in driving the growth and sustainability of tribal tourism. These findings align closely with existing scholarship, which emphasizes the importance of empowering local communities and leveraging cultural assets to create authentic and enriching tourism experiences (Suresh, 2021; Telang & Bokde, 2019).

The second factor, termed "Traditional Aspect of Tribal Society," highlights the significance of tribal tourism in preserving indigenous cultures, facilitating cultural exchange, and conserving heritage. Tribal communities view tourism as a means of safeguarding their traditional practices, rituals, and languages, while also sharing their unique heritage with visitors in a respectful and meaningful manner. This finding resonates with the broader literature, which underscores the transformative potential of tourism in fostering cross-cultural understanding and preserving intangible cultural heritage (Arora & Sharma, 2022; Smith, 1996).

The third factor, denoted as "Sustainable Development in Tribal Area," underscores the imperative of infrastructure development, accessibility, hospitality services, and sustainable practices for the holistic development of tribal tourism. These findings echo the growing recognition of sustainable tourism as a key driver of socio-economic progress and environmental conservation in tribal areas. The emphasis placed on sustainable development aligns with the principles of responsible tourism, which prioritize the well-being of local communities and the preservation of natural and cultural resources (Chen, 2019; Verma & Murdia, 2017).

Comparing our findings with the broader literature on tribal tourism, several noteworthy parallels and divergences emerge, shedding light on both the successes and challenges facing tribal tourism initiatives. While our study corroborates existing research on the positive impacts of tribal tourism on community empowerment, cultural preservation, and sustainable development, it also highlights areas for improvement and further exploration.

For instance, our findings underscore the need for greater community involvement in decision-making processes and the equitable distribution of tourism benefits. Addressing concerns related to cultural commodification, environmental degradation, and socio-economic disparities requires a concerted effort from policymakers, practitioners, and stakeholders to foster inclusive, sustainable, and culturally sensitive tourism initiatives in tribal areas.

In conclusion, our study contributes valuable empirical insights to the ongoing discourse on tribal tourism by elucidating tribal perceptions towards tourism development. By identifying key factors influencing tribal attitudes and priorities, our findings offer actionable implications for policymakers and practitioners seeking to design and implement effective strategies for promoting inclusive, sustainable, and culturally sensitive tourism initiatives in tribal areas.

Suggestions towards Development of Tribal Tourism

The following are some proposals that can be put into practice to help enhance tribal tourism. The first point to remember is that all aspects of tribal tourism must be done in compliance with the rights of the tribes. The UN declaration provides certain rights such as these -Territorial rights, Bio-cultural rights, Spiritual rights, Decision-making rights and Free prior and informed consent.

We can deduce from these rights that they are, first and foremost, the original proprietors of the property, and that obtaining permission from them and refraining from making any alterations until they agree is necessary. They have the right to teach and maintain the special relationship they share with their land, resources, and water in any way they want, without

Secondly, as suggested is the WINTA's ITEF (Indigenous Tourism engagement framework). It involves six integrated guiding principles of respect, consult, empower, partner, benefit, and protect are derived from the UN Declaration on the Rights of Indigenous Peoples (UNDRIP) and the Larrakia Declaration, and they have been compared to other important international guidelines on Indigenous peoples' rights. The arrow on the left represents a broad group of major stakeholders in Indigenous tourism. It identifies the various stakeholders as the governments, supporting bodies such as NGOs or volunteers may form a part of this, the tourism industry, Individuals and organizations within these interconnected bodies will be more likely to participate in friendly and mutual engagements if they unite together and adopt the Larrakia principles in their interactions.

Secondly a framework has been proposed which is known as WINTA's tourism framework. As per this framework, it is suggested that the public sector in collaboration with the tribal community could work together to produce unique tourism resources which would form a basis for the way the trend would change in the coming future as well as for development purposes, while creating a comprehensive strategy which would connect numerous tribal features and picturesque areas.

Case Study of Kinnaur

Kinnaur is a well-known district in the picturesque state of Himachal Pradesh, which comprises all in all twelve districts and it is one of them. Its administrative headquarters are located in Reckong Peo. It is the second least populous district with 84,121 people residing there following Lahaul and Spiti in Himachal Pradesh as per the 2011 population census. The state of Himachal Pradesh consists of 68.65 lakh population of tribes as per 2011 census and Kinnaur has 48,746 of them. As per the official statistics of 2011–12 published by the Ministry of Tourism, it has been published that 17,904,702 domestic visitors and 1,090,598 international visitors visited Kinnaur.

The Tribe which resides in Kinnaur is referred to as the Kinnaura Tribe which comprises the largest ethnic group in the area. This particular location was once thought to be a forbidden zone, but it is today a popular tourist attraction, particularly for adventure and mystery seekers. Tourism has been steadily growing in the area, with numerous new hotels and homestays springing up to meet the demands of visitors. Residents reported that they are not opposed to development of tourism; rather, they were in supportive of additional tourism development and saw a very minimal impact. According to the findings, locals enjoy the presence of foreigners in their communities, and tourists do not pose a threat to their existence. Furthermore, residents reported that their understanding of the environment and its management had increased.

In terms of challenges, if the appropriate government support is provided the advancement opportunities would flourish more than ever along with focusing on other aspects such as illiteracy, lack of awareness and accessibility which need to be taken up so integrated development takes place. Most of the respondents have emphasized upon the fact that the major socio-economic benefits which are derived from tourism are important to their economy in terms of revenue generation and also diversifying the sources of income generation. The community feels that in these current times everything comes at a place and due to which tourism would highly enhance them to be self-sufficient. It has been focused that in comparison to urbanization and globalization the impact of tourism is extremely minimal. They have currently improved widely in terms of infrastructure, socio-economic benefits, multiple sources of income and cultural exchange.

Conclusion and Recommendations

It can be inferred through the study that tribal tourism has a lot of potential to expand in our country due to the large number of tribal people. The majority of the concerns raised by my survey results are similar to those identified in the literature review, including a lack of government support, accessibility issues, a lack of awareness leading to resistance, and illiteracy among the tribes. The advantages of tourism can be maximized, and many negative consequences can be avoided if they are well managed. All these would ultimately lead to them becoming self-sustaining, empowered and the integrated development of every individual in our society.

The recommendations which can be suggested towards the development of tribal tourism in sync with four Hs of tribal tourism are as follows:

- Informed consent from the community concerned through effective communication.
- Respecting the wishes of tribal population on whether they wish to engage in tourism-related activities.
- Government intervention to mandate policies and ensure effective regulation of these policies on how tribal tourism is to take place and no negative consequences occur.
- Tours being led only by certified travel agencies which involve responsible travel.
- Ensure all the benefits are reaching directly to the community.
- Monitor Tribal Community-Tourist interaction so that only positive exchange occurs.
- Introducing the option of permits thus a thorough check of the person which may include information about them, their purpose of visit and health check so no possible risk is there to the community.
- Time to time interaction with the tribes to understand them, focus and solve their challenges or issues arising, focus on literacy and awareness of the tribes.
- Increasing the range of accessibility to these regions with monitoring and determination of carrying capacity in all three forms- Environment, Economic and Social.

References

Arora, S., & Sharma, A. (2022). *Event tourism in Asian countries*. CRC Press.

Barelli, M. (2012). Free, prior and informed consent in the aftermath of the UN declaration on the rights of indigenous peoples: Developments and challenges ahead. *The International Journal of Human Rights*, *16*(1), 1–24. https://doi.org/10.1080/13642987.2011.597746

Bott, E. (2018). Among the piranhas: The troubling lifespan of ethnic tropes in 'tribal' tourism to Vietnam. *Journal of Sustainable Tourism*, *26*(8), 1291–1307. https://doi.org/10.1080/09669582.2018.1435669

Buckley, R. (2012). Sustainable tourism: Research and reality. *Annals of Tourism Research*, *39*(2), 528–546. https://doi.org/10.1016/j.annals.2012.02.003

Butler, R. W., & Pearce, D. G. (2003). *Contemporary issues in tourism development*. [online] *Google Books*. Routledge. Available at https://books.google.co.uk/books?hl=en&lr=&id=oNSFAgAAQBAJ&oi=fnd&pg=PA189&dq=four+Hs+of+tribal+tourism&ots=bY5dfvnx11&sig=HPMPsbV5APUlL371kGv3CHHMX8A&redir_esc=y#v=onepage&q&f=false [Accessed 29 Oct. 2023].

Caplan, A. L., & Moreno, J. D. (2011). The Havasu 'Baaja tribe and informed consent. *The Lancet*, *377*(9766), 621–622. https://doi.org/10.1016/s0140-6736(10)60818-5

Chang, H.-M., Hung, C.-H., & Chou, C.-L. (2018). A study of Alishan indigenous tribal tourism development in Taiwan. *Asian Journal of Environment & Ecology*, *8*(1), 1–12. https://doi.org/10.9734/ajee/2018/45394

Chang, H.-M., Lin, C.-H., & Chuang, Y.-W. (2021). Study on sustainable development strategy of indigenous tribal tourism. *E3S Web of Conferences*, *257*, 03045. https://doi.org/10.1051/e3sconf/202125703045

Chen, H.-S. (2019). Establishment and application of an evaluation model for Orchid Island sustainable tourism development. *International Journal of Environmental Research and Public Health*, *16*(5), 755. https://doi.org/10.3390/ijerph16050755

Chouhan, V. (2022). Developing a sustainable tribal tourism model vis-a-vis the tribal region of Rajasthan. *Heritage & Services Marketing Suggested Citation*, *8*(1), 58–63. https://doi.org/10.5281/zenodo.6594696

Chu, J. (2015). *ROLES OF ABORIGINAL TOUR GUIDES IN ABORIGINAL TRIBAL TOURISM: CULTURAL IDENTITY AND ETHNIC RECOGNITION - ProQuest*. [online] www.proquest.com. Available at https://www.proquest.com/openview/caf4c61983c8fd16dbf1622cef0a0e3e/1?pq-origsite=gscholar&cbl=626342 [Accessed 29 Oct. 2023].

Doğan, M., & Timothy, D. J. (2019). Beyond tourism and taxes: The ecomuseum and social development in The Ak-Chin tribal community. *Journal of Tourism and Cultural Change*, *18*(2), 133–149. https://doi.org/10.1080/14766825.2019.1593994

Dyer, P., Aberdeen, L., & Schuler, S. (2003). Tourism impacts on an Australian indigenous community: A Djabugay case study. *Tourism Management*, *24*(1), 83–95. https://doi.org/10.1016/s0261-5177(02)00049-3

Fitch, R. A., Mueller, J. M., Ruiz, R., & Rousse, W. (2022). Recreation matters: Estimating millennials' preferences for native American cultural tourism. *Sustainability*, *14*(18), 11513. https://doi.org/10.3390/su141811513

Fredericks, C. F. (2016). Operationalizing free, prior, and informed consent, *Albany Law Review*, *80*, 429. Available at https://heinonline.org/HOL/LandingPage?handle=hein.journals/albany80&div=18&id=&page= [Accessed 16 Dec. 2020].

Goeldner, C. R., & Gartner, W. C. (1995). Conference report. *Journal of Travel Research*, *34*(2), 62–65. https://doi.org/10.1177/004728759503400212

Hair, J. F., Risher, J. J., Sarstedt, M., & Ringle, C. M. (2019). When to use and how to report the results of PLS-SEM. *European Business Review*, *31*(1), 2–24.

Huang, H.-C., Chia-Hao, L., & Chang, H.-M. (2015). Does tourism development bring positive benefit to indigenous tribe? Case by Dongpu in Taiwan. *Advances in Research*, *4*(4), 235–246. https://doi.org/10.9734/air/2015/15030

Ishii, K. (2012). The impact of ethnic tourism on hill tribes in Thailand. *Annals of Tourism Research*, *39*(1), 290–310. https://doi.org/10.1016/j.annals.2011.05.004

Johnston, A. M. (2014). *Is the sacred for sale*. Routledge. https://doi.org/10.4324/9781849773874

Kumar, J., & M. (2016). Tribal tourism for sustainable development in Himachal Pradesh: Issues and opportunities. (4th ed., Vol. 1). Avahan: A Journal on Hospitality and Tourism.

Malhotra, N. K. (2007). Review of marketing research. *Review of Marketing Research*, v–v. https://doi.org/10.1108/s1548-6435(2007)0000003004

Maraña, E.C., Chantyl G. Aveno, Magtibay, M., Mae, A. and Iriman, M. M. (2023). Dumagat tribes and tourism industry engagement in promoting cultural integrity. *18*(2), pp. 180–196. https://doi.org/10.30574/wjarr.2023.18.2.0804

McGregor, J., & Tsosie, R. (2021). Rethinking research protections for tribal communities. *The American Journal of Bioethics*, *21*(10), 30–32. https://doi.org/10.1080/15265161.2021.1965244

Musavengane, R. (2019). Using the systemic-resilience thinking approach to enhance participatory collaborative management of natural resources in tribal communities:

Toward inclusive land reform-led outdoor tourism. *Journal of Outdoor Recreation and Tourism*, *25*, 45–56. https://doi.org/10.1016/j.jort.2018.12.002

Notzke, C. (2004). Indigenous tourism development in Southern Alberta, Canada: Tentative engagement. *Journal of Sustainable Tourism*, *12*(1), 29–54. https://doi.org/10.1080/09669580408667223

Perrault, A., Herbertson, K., & Lynch, O. J. (2006). Partnerships for success in protected areas: The public interest and local community rights to prior informed consent (PIC), *Georgetown International Environmental Law Review*, *19*, 475. Available at https://heinonline.org/HOL/LandingPage?handle=hein.journals/gintenlr19&div=22&id=&page= [Accessed 29 Oct. 2023].

Pratt, S., Gibson, D., & Movono, A. (2013). Tribal tourism in Fiji: An application and extension of Smith's 4Hs of indigenous tourism. *Asia Pacific Journal of Tourism Research*, *18*(8), 894–912. https://doi.org/10.1080/10941665.2012.717957

Rao, R. (2016). Informed consent, body property, and self-sovereignty. *The Journal of Law, Medicine & Ethics: A Journal of the American Society of Law, Medicine & Ethics*, *44*(3), 437–444. https://doi.org/10.1177/1073110516667940.

Sengupta, C. (2018). *Government intervention and the future of tourism in India: Analysis of the Swadesh Darshan Scheme* (2nd ed.). Perspectives, Patterns and Practices.

Sharma, A. (2023). *Indigenous and tribal tourism*. [online] www.igi-global.com. Available at: https://www.igi-global.com/chapter/indigenous-and-tribal-tourism/322422 [Accessed 29 Oct. 2023].

Sibi, D. P., & Anjaneya Swamy, G. (2015). Scope and potential for indigenous tourism – An analytical study in Wayanad, Kerala. *ZENITH International Journal of Multidisciplinary Research*, *5*(11), 2231–5780. http://zenithresearch.org.in/images/stories/pdf/2015/NOV/ZIJMR/7_ZIJMR_VOL5_ISSUE11_NOV2015.pdf

Smith, V. L. (1996). The four Hs of tribal tourism: Acoma - a pueblo case study. *Progress in Tourism and Hospitality Research*, *2*(3–4), 295–306. https://doi.org/10.1002/pth.6070020310

Srivastava, S. (2021). *Capacity building through heritage tourism: An international perspective*. [online] *Google Books*. CRC Press. Available at https://books.google.co.uk/books?hl=en&lr=&id=_dIREAAAQBAJ&oi=fnd&pg=PA171&dq=four+H+of+tribal+tourism&ots=vzQpNge3KX&sig=LrR8Q9gqT_LSK-AmobNBWeWURA4&redir_esc=y#v=onepage&q=four%20H%20of%20tribal%20tourism&f=false [Accessed 29 Oct. 2023].

Tabachnick, B. G., & Fidell, L. S. (2001). *Using multivariate statistics*. Allyn & Bacon.

Telang, D., & Bokde, S. (2019). *The role and importance of tribal tourism in the sustainable development of tribal communities in Palghar district* (5th ed.). Research Journey.

Verma, S., & Murdia, M. (2017). Highlighting tribal tourism potentials of Southern Rajasthan. *IRA-International Journal of Management & Social Sciences (ISSN 2455-2267)*, *6*(3), 384. https://doi.org/10.21013/jmss.v6.n3.p5

Whyte, K. P., Selinger, E., & Outterson, K. (2011). Poverty tourism and the problem of consent. *Journal of Global Ethics*, *7*(3), 337–348. https://doi.org/10.1080/17449626.2011.635689

Wixom, B. H., & Todd, P. A. (2005). A theoretical integration of user satisfaction and technology acceptance. *Information Systems Research*, *16*(1), 85–102. https://doi.org/10.1287/isre.1050.0042

8

BRANDING AND REBRANDING DESTINATIONS THROUGH HERITAGE TOURISM

Dallen J. Timothy

Introduction

Cultural heritage has been in demand among tourists for centuries, and today heritage-based tourism is one of the largest segments of the tourism industry. Countries, regions, and localities become popular tourist destinations largely because of their heritage assets and cultural resources. Many destinations boast of world-class historic monuments that appeal to tourists on a mass global scale, whereas most of the world's heritage features are national, regional, or local in their scale and market reach (Timothy, 1997). The tourism industry in some destinations is nearly synonymous with heritage tourism, such as in Egypt, China, Cambodia, and Peru, where other important types of tourism (e.g., cruises, beach resorts, and nature-based tourism) regularly pale compared to the global prominence of those countries' heritage sites.

Expressions of tangible and intangible heritage are frequently utilized by governors to build a sense of national solidarity, such as through depictions on currency and stamps, or to tap into the educational and scientific value of heritage through educational curricula. From a tourism perspective, these same cultural features are often utilized for marketing purposes, often entrenched in campaign slogans, symbols, logos and taglines (Seraphin et al., 2018; Tsaur et al., 2020). Because of their universal appeal, heritage symbols feature prominently on websites and in tourist brochures and guidebooks. For instance, India's hugely successful 'Incredible India' campaign has featured the Taj Mahal many times over, with that momentous site often symbolizing the country's heritagescape within the global tourist gaze (Bhakuni & Kapoor, 2011). Emphasizing the role of cultural heritage in a destination's tourism milieu is a form of branding. This chapter

DOI: 10.4324/9781003346197-8

examines the notion of heritage as a tourism brand. It considers the role of heritage in place branding from a general perspective and then hones in on particular international and national heritage brands that are often believed to hold tourism value. Finally, the chapter examines the notion of heritage as a mechanism for recovery branding in times of crisis or when other economic sectors fail.

Heritage Place Branding

Place (destination) branding has been a significant focus of destination management organizations (DMOs), economic development specialists, government agencies, and tourism scholars for many years (Foroudi et al., 2020; Morgan et al., 2004). Destination branding generally refers to a process of identifying, shaping, and fostering the inimitable identity of a place. In recent decades, this has entailed deliberate actions (e.g., destination marketing and policy-making) to promote a destination's unique qualities, highlighting its tourism assets, services, activities, and place-based characteristics to create a reputation and underscore its value as a tourist destination. Although destination branding and marketing are intentional actions, a destination's image, or brand, may also develop organically through time as the place becomes closely associated with certain iconic natural or cultural characteristics that stand out from those in other places. For example, the Pyramids of Giza, the Temples of Luxor, and other prominent archaeological sites have epitomized Egyptian tourism for centuries (El Daly, 2016; Mustafa, 2021; Timothy & Tahan, 2020).

Heritage is frequently the central feature of both purposive and organic place branding for several reasons. Among many others, heritage provides a historical context for today's societies. It is rooted in place and illuminates how places and social identities have developed over time. Heritage branding also highlights the living culture of destinations in terms of cuisine, faith, festivals and celebrations, languages, social mores, and cultural traditions. Organically speaking, these heritage characteristics work together to create images of destinations that are known for their unique architectural traditions, cultural landscapes, and living practices. Purposive branding regularly uses these same features to create a unique selling point and competitive advantage for tourist destinations. In fact, perhaps more than any other asset, heritage is widely used to distinguish destinations worldwide through brand recognition, brand positioning and brand identity (Haq et al., 2021; Hussin, 2018; Kim et al., 2022; Saeedi & Heidarzadeh Hanzaee, 2018).

Heritage-based branding has achieved considerable success in many locations owing to heritage's extreme tourism value (Nazir et al., 2023; Timothy & Boyd, 2003, 2006). Together with its corresponding concepts—authenticity, nostalgia, preservation, and education—heritage is valorized as a set of precepts that draw people for many reasons, including education, curiosity, self-seeking, furthering a hobby, religious obligations, experience unmatched adventures, and more (Timothy, 2021).

A salient part of this image-making and visibility-raising trend is the establishment of official heritage registers and the inscription of cultural sites and places on them. These listings often add brand appeal and prestige to existing heritage sites and can be seen at many different levels, including international, national, and local. Many countries have established heritage designations that project a certain level of repute and raise the place's tourism value. In the United States (USA), for example, the National Park Service maintains the National Register of Historic Places and manages the National Historic Landmarks program—both of which are coveted trademarks in the heritagescape of the USA (Cullingworth, 1992; Maskey et al., 2009). On the global scale, UNESCO's World Heritage List is the best-known and most prestigious brands, although more recent designations also include UNESCO's Intangible Cultural Heritage and Cultural Landscapes. UNESCO's seal of approval is now commonly regarded by many countries (and some tourists) as a 'tourism logo' that manifests a certain level of authenticity, protection, and quality (Hamid et al., 2018; King & Halpenny, 2014).

International Heritage Brands

UNESCO World Heritage Sites: The Global Heritage Brand

The most recognized global heritage brand is UNESCO's World Heritage Sites (WHSs) (Adie, 2017; Ryan & Silvanto, 2009; Timothy, 2021). In 1972, UNESCO established the Convention Concerning the Protection of the World Cultural and Natural Heritage, which came into force in 1975. With the exception of Liechtenstein and Nauru, all of the world's sovereign states (195) have ratified the treaty, which aims to spotlight and protect cultural and natural sites that exemplify humankind's accomplishments and Earth's natural beauty, and which demonstrate universal heritage value. At the close of 2023, there were 1,199 UNESCO World Heritage Sites in 168 signatory countries, including a wide range of natural, cultural, and mixed heritage locales.

Although the World Heritage Site designation was not created to promote tourism, it has emerged as a desirable brand that countries use to position themselves in the tourism marketplace (Adie, 2017; Hamid et al., 2018; Jiménez Madariaga & Seño Asencio, 2019; Mota et al., 2017; Ryan, 2015; Santamarina, 2023). Some countries are covetous of the UNESCO brand and attempt to inscribe as many heritage places as possible on the global list, ostensibly (and perhaps misguidedly) for the brand value in attracting tourists. The assumption that the WHS brand increases visitation, however, is open to debate and in need of additional research. Some studies suggest that the WHS brand is a major tourist draw and results in increased visitation (De Simone et al., 2019), although the majority of observers concede that the listing itself does not inevitably result in increased visitation (Adie & Hall, 2017; Iaţu et al., 2018; Mota et al., 2017). Instead, it is more generally accepted that variables such as accessibility; pre-WHS listing popularity (or lack thereof); community empowerment, involvement, capacity-building, solidarity and sense of pride; political stability; and marketing efforts are more likely to contribute to tourism growth than the actual WHS brand (Amer et al., 2023; Hall & Piggin, 2002; Hernández, 2019; Jimura, 2016; Ryan & Silvanto, 2010; Shabani et al., 2021; Timothy, 2018).

Despite its growing popularity, the UNESCO WHS brand is not without controversy. Some scholars have argued that WHS inscription is highly political with nepotism, favoritism, and even corruption potentially occurring in the preparation and nomination process (Meskell et al., 2015). Meskell (2018) also argues that regional imbalances and Eurocentrism are endemic to the traditions of UNESCO's World Heritage List. Other commentators contend that World Heritage-branded sites may overshadow other historic areas within a country, creating imbalances and endangering other forms of cultural expressions (Amer et al., 2023) that may be relatively neglected by heritage protection agencies as they devote more energy and resources to WHSs. Regardless of these critiques, the WHS brand continues to be an important foundation of many heritage destinations' promotional efforts.

European Capitals of Culture

Another of the best-known heritage brands at the international level is the European Capital of Culture. Each year, one or more historic cities are selected to become a future Capital of Culture. This initiative by the European Commission was launched in 1985, with Athens being designated the first European City of Culture, followed by Florence in 1986 and Amsterdam in 1987. After 1999, the European City of Culture scheme was retitled the European Capital of Culture (Patel, 2013). This designated has been quite inclusive with 68 cities from the majority of Europe's countries having received the prestigious title by 2023 (Table 8.1). Until 1999, only one city had been crowned a Capital of Culture per year, but in 2000 nine cities were selected, and in the years following, two or three cities have received the honorific of being a European Capital of Culture.

Only countries that are members of the European Union or candidate and potential EU member states (i.e., Albania, Bosnia and Herzegovina, Kosovo,

TABLE 8.1 European Capitals of Culture: Past, Present, and Future

Year	*City/Cities*	*Country/Countries*
1985	Athens	Greece
1986	Florence	Italy
1987	Amsterdam	Netherlands
1988	West Berlin	West Germany
1989	Paris	France
1990	Glasgow	United Kingdom
1991	Dublin	Ireland
1992	Madrid	Spain
1993	Antwerp	Belgium
1994	Lisbon	Portugal
1995	Luxembourg	Luxembourg
1996	Copenhagen	Denmark
1997	Thessaloniki	Greece
1998	Stockholm	Sweden
1999	Weimar	Germany
2000	Avignon	France
	Bergen	Norway
	Bologna	Italy
	Brussels	Belgium
	Helsinki	Finland
	Kraków	Poland
	Prague	Czech Republic
	Reykjavík	Iceland
	Santiago de Compostela	Spain
2001	Rotterdam	Netherlands
	Porto	Portugal
2002	Bruges	Belgium
	Salamanca	Spain
2003	Graz	Austria
2004	Geno	Italy
	Lille	France
2005	Cork	Ireland
2006	Patras	Greece
2007	Sibiu	Romania
	Luxembourg	Luxembourg
2008	Liverpool	United Kingdom
	Stavanger	Norway
2009	Linz	Austria
	Vilnius	Lithuania
2010	Essen	Germany
	Istanbul	Turkey
	Pécs	Hungary
2011	Tallinn	Estonia
	Turku	Finland

(Continued)

TABLE 8.1 (Continued)

Year	*City/Cities*	*Country/Countries*
2012	Guimarães	Portugal
	Maribor	Slovenia
2013	Košice	Slovakia
	Marseille	France
2014	Riga	Latvia
	Umeå	Sweden
2015	Mons	Belgium
	Plzeň	Czech Republic
2016	San Sebastián	Spain
	Wrocław	Poland
2017	Aarhus	Denmark
	Paphos	Cyprus
2018	Leeuwarden	Netherlands
	Valletta	Malta
2019	Matera	Italy
	Plovdiv	Bulgaria
2020/2021	Galway	Ireland
	Rijeka	Croatia
2022	Esch-sur-Alzette	Luxembourg
	Kaunas	Lithuania
	Novi Sad	Serbia
2023	Elefsina	Greece
	Timişoara	Romania
	Veszprém	Hungary
2024	Bad Ischl	Austria
	Bodø	Norway
	Tartu	Estonia
2025	Chemnitz	Germany
	Nova Gorica/Gorizia	Slovenia/Italy (border city, shared nomination)
2026	Oulu	Finland
	Trenčin	Slovakia
2027	Évora	Portugal
	Liepāja	Latvia
2028	České Budějovice	Czech Republic
	Bourges	France
	Skopje	North Macedonia

Source: European Commission (2023).

Montenegro, North Macedonia, Serbia, and Turkey), as well as non-EU members of the European Free Trade Association (EFTA) (i.e., Iceland, Liechtenstein, Norway, and Switzerland) are eligible to host a Capital of Culture. Because it left the European Union in 2020, the United Kingdom is no longer eligible (European Commission, 2023).

This designation aims to bring visibility to cities throughout Europe, assist with urban renewal, grow tourism, celebrate a pan-European identity and common heritage, and revel in the uniqueness of each locality with a year full of events, performances, and other art and culture appreciation activities (Patel, 2013). Despite the brand officially being designated for only one year, in practice, "once a Capital of Culture, always a Capital of Culture". Thus, in common with Olympic host cities (Herstein & Berger, 2013; Zhang & Zhao, 2009), the promotional, place-making, and branding efforts undertaken in the years preceding a city's designated year essentially brand a city with long-term visibility and an international profile. This can effectively enhance a destination's community pride and build solidarity among its inhabitants (European Commission, 2023).

Three European Capitals of Culture were designated for 2023: Elefsina, Greece; Veszprém, Hungary; and Timisoara, Romania. Each year's Capitals of Culture have been announced through 2028 (Table 8.1), giving each of the future cultural centers a head-start opportunity to prepare for the events of their assigned year and to create their unique heritage brand.

Council of Europe's Cultural Routes Program

Another prestigious European brand is the Cultural Routes of the Council of Europe. This program began in 1987 to link European places together, highlighting a shared pan-European heritage, and distributing tourism's economic impacts to diverse localities throughout the continent. The Cultural Routes are designated to promote the values of the Council of Europe, including cultural democracy, human rights, mutual understanding, intercultural dialogue, cross-border collaboration, and a greater understanding of European history (Council of Europe, 2023).

The Cultural Routes are long-distance, transnational networks that link together individual or collective sites into thematic itineraries that can be traveled in part by foot or entirely by car or public transportation (Timothy & Boyd, 2015). In 2023 there were 47 designated Cultural Routes traversing the territories of nearly all countries of Europe. Some key thematic routes include the Iron Curtain Trail, the European Fairy Tale Route, European Route of Industrial Heritage, Routes of the Olive Tree, and the European Route of Jewish Heritage. Implementing principles such as clustering and critical mass, these purposely designated networks promote comparable and related sites in a way that is more impactful than any of the individual nodes acting on its own as a heritage attraction (Timothy & Boyd, 2015). For a locality to be included as part of a cultural route lends significant brand value and prestige, and reinforces its role in upholding the values of the European Commission (Severo, 2017).

Individual National Heritage Brands

Many countries have established their own heritage brand listings, which often elevates social value, educational importance, and authenticity while strengthening national solidarity. Some countries have several different and often overlapping brands. For example, Japan's Agency for Cultural Affairs maintains a list of Cultural Properties, whereas the country's 'National Treasures', also designated by the Agency for Cultural Affairs, are the most precious of Japan's tangible cultural properties (Jimura, 2022). Sometimes these listings are also esteemed for their touristic value. Table 8.2 presents several examples of national-level heritage designations that lend value and prestige to their heritage brands.

TABLE 8.2 Examples of National Heritage Listings and 'Brands'

*Listing**	*Brand*	*Country*	*Number of Designated Heritage Sites/Areas*	*Designating Authority*
Australian National Heritage List	National Heritage Place	Australia	ca 120	Dept of Climate Change, Energy, the Environment and Water
National Historic Sites of Canada	National Historic Site	Canada	ca 1,005	Minister of the Environment
National Monuments	National Heritage Sites	Chile	ca 940	Council of National Monuments of Chile
Major Historical and Cultural Site Protected at the National Level	Nationally Protected Sites	China	ca 5,060	State Administration of Cultural Heritage
National Register of Sites and Monuments	Sites and Monuments	Denmark	ca 165,000	Agency for Culture and Palaces
National Register of Cultural Monuments	Cultural Monuments	Estonia	ca 26,497	National Heritage Board of Estonia
Monuments of National Importance	Monument of National Importance	India	ca 3,684	Archaeological Survey of India and Union Government of India
Cultural Properties	National Treasure, Important Cultural Property & Historic Site	Japan	16,000 + (national-level properties)	Agency for Cultural Affairs

(Continued)

TABLE 8.2 (Continued)

*Listing**	*Brand*	*Country*	*Number of Designated Heritage Sites/Areas*	*Designating Authority*
National Inventory of Cultural Heritage	Historic Monument and Site	Morocco	ca 398	National Institute of Archaeology and Heritage Sciences
List of Cultural Monuments in Norway	Cultural Monument	Norway	115,000 +	National Archives
List of National Cultural Treasures	National Cultural Treasure	Philippines	ca 106	National Commission for Culture and the Arts
National Cultural Heritage Register	Monuments of Culture and History	Russia	100,000 +	Federal Service for Monitoring Compliance with Cultural Heritage Legislation
List of Cultural Heritage of Exceptional Importance	Cultural Monuments, Archaeological Sites, Historic Landmark & Cultural-Historical Unit of Exceptional Importance	Serbia	ca 2,592	National Assembly of Serbia
Swiss Inventory of Cultural Property of National and Regional Significance	Cultural Property of National Significance	Switzerland	ca 2,416	Federal Office of Civil Protection
National Historic Sites of Tanzania	National Historic Site	Tanzania	60 +	Ministry of Natural Resources and Tourism
State Register of Immovable Monuments	Immobile Landmark	Ukraine	ca 3,300	Ministry of Culture
List of National Historic Landmarks	National Historic Landmark	United States	ca 2,500	National Park Service
Scheduled Monuments	Scheduled Ancient Monument	Wales (UK)	ca 4,186	Cadw
National Monuments of Zimbabwe	National Monument	Zimbabwe	ca 170	National Museums and Monuments of Zimbabwe

* Note: Some countries have more than one official heritage register.

Source: Compiled by the author from multiple sources.

China's 'A' Tourist Attraction Rating System

In an effort to grow tourism, China established an attraction grading system in 1999 known as the Tourist Attraction Rating Categories, which esteems cultural and natural attractions according to their national value and importance, safety, accessibility, and other variables. There are five rankings: A, AA, AAA, AAAA, and AAAAA. The 5-A ranking is the most prestigious designation and is assigned to the country's most magnificent attractions (Lin et al., 2020; Ryan et al., 2009). This program is administered by China's Ministry of Culture and Tourism and is a hugely successful effort to brand the country's best attractions. It has found most success among domestic Chinese visitors, but the 5-A and 4-A attractions also tend to wield significant international appeal. Examples of 5-A attractions include the Longmen Grottoes in Luoyang, the Forbidden City in Beijing, and the Terracotta Warriors in Xi'an. A high A-rank is a mark of distinction that many cities and provinces in China aspire to receive, and research shows that high-level ratings in this system result in greater competitive advantages and significant tourism growth (Ryan et al., 2009; Wang et al., 2020).

The Magical Towns of Mexico

Pueblos Mágicos (magical towns) are one of the most coveted heritage brands in Mexico. The Magical Towns Program was initiated in 2001 by Mexico's Secretary of Tourism to alleviate poverty by spreading tourism to small towns and villages throughout the country, and to help protect Mexico's small-town heritage (Flores-Romero et al., 2021). This recognition showcases towns' and villages' 'magical' attributes, such as exceptional histories, beautiful built and natural environments, storied legends, authentic traditions, special events and celebrations, unique culinary heritages, distinct handicrafts and art, and high-quality hospitality (Olivé & Torres-González, 2023; Secretaría de Turismo, 2016). According to Journey Mexico (2023, n.p.), the Pueblo Mágico brand is a "mark of excellence" and signals "that the town is worth exploring for a day or two".

By the end of 2023, there were 177 Pueblos Mágicos, with at least one in every state in the country outside of Mexico City (Journey Mexico, 2023). To qualify for this prestigious identifier, a town or village must have a population over 5,000 and should be located near enough to a larger city to support a tourism market base. A local committee applies for the designation to the Secretary of Tourism and must demonstrate the community's ability and willingness to fund and plan tourism successfully. After receiving the prestigious brand, the community must continue to maintain its magical qualities and focus on tourism-based economic development (Secretaría de Turismo, 2016). The Pueblo Mágico brand is a significant point of pride for community residents, leaders, and entrepreneurs, as it denotes a level of specialness and Mexicanness that not every village and town in Mexico can demonstrate.

Inspired by Mexico's village branding program, a similar series of Pueblos Mágicos was established in Ecuador in 2019 by that country's Ministry of Tourism. In 2023, there were 21 magical towns that are deemed to epitomize Ecuador's best cultural heritage features. According to the country's Ministry of Tourism (2023, n.p.), "Ecuador is infinitely diverse and our four worlds are a paradise of life and heritage, capable of making your spirit vibrate with magical adventures. On this occasion, we want to introduce you to 21 fascinating locations that await you in the most beautiful corners with unique stories and traditions, surprising natural spaces, ancestral cultures, exquisite gastronomy and very charismatic and talented people".

United States Heritage Lists

The USA has several heritage brands, many of which overlap and coincide with one another. One of the earliest heritage brands in the USA, National Historic Landmarks (NHLs), began with heritage protective legislation in 1935 but was initiated officially in 1960. That year, 93 cultural sites were designated NHLs (Frank & Petersen, 2002), and today there are more than 2,600 NHLs throughout the country and its overseas territories. This program is administered by the Department of the Interior's National Park Service and includes historic buildings, structures, sites, objects and districts that illustrate the heritage of the USA (National Park Service, 2023a). The NHL brand holds social and economic value for communities around the country and is granted based upon the following principles (Martens & Timothy, 2006):

- Possesses national significance.
- Represents the heritage of the USA in history, architecture, archeology, engineering, and culture.
- Demonstrates a high degree of integrity of location, design, setting, materials, workmanship, feeling, and association.
- Maintains authenticity.
- Adheres to various site-specific criteria.

One particular example of the value of the NHL brand is the historic town of Tombstone, Arizona. Founded in 1879, Tombstone became a thriving silver mining community in the late 1800s with a peak population of around 14,000, but after technical challenges in the mining sector, by 1910, the village's population had dwindled to only a few hundred. During its heyday, Tombstone was made famous by the 1881 Gunfight at the O.K. Corral, which ensued between law enforcement and a gang of outlaws. This brief event is arguable one of the most famous shootouts to have occurred in the old Wild West and is the most salient element of the town's tourism appeal today (McCormack, 2016).

Known for its Wild West history of mining and lawlessness, Tombstone is one of Arizona's top heritage destinations, with more than 400,000 people visiting each year to experience the nineteenth-century ambience and daily reenactments of the famous shootout. Several books have been written and movies made about Tombstone, adding additional notoriety and visibility to the destination (Boggs, 2019).

The Tombstone Historic District was inscribed a National Historic Landmark in 1962 based upon the criteria noted above. However, after years of evaluating the town's NHL status, the Department of the Interior classified Tombstone as 'threatened' in 2004, meaning that the historic district had suffered, or was in imminent danger of suffering, a severe loss of historical integrity, which meant a potential loss of NHL branding (City of Tombstone, 2012). Since its 1972 historic preservation plan, which determined 1885 as the baseline year for preservation and restoration, the historic district has undergone significant changes that undermined the area's authenticity. These included, but were not limited to, inappropriate building materials, inaccurate historic dates on buildings, replacing historic features rather than repairing the originals, incompatible new builds and repairs in the historic district, integrating architectural and frontage pieces that would not have existed at the time, and modern encroachments such as illuminated signage on some downtown businesses (Martens & Timothy, 2006). The endangerment of Tombstone's listing was of grave concern to community and business leaders, as it was felt to be a direct threat to the town's tourism sector.

After additional consultations with the National Park Service and hired consultants, a new plan was adopted in 2005 to address these issues. The old historic preservation strategy was revised and updated, and new preservation guidelines were established. A new historical integrity timeframe was established, extending the 1885 baseline to 1931, because this was when the town decided to resurrect its economy by focusing on tourism. The historic area was re-defined, and the historic preservation zones were expanded. As a result of these actions and others, the National Park Service elevated the Tombstone National Historic Landmark from threatened to 'watch', which still requires regular monitoring (City of Tombstone, 2012).

Another list maintained by the US National Park Service is the National Register of Historic Places, which identifies historic places deemed worthy of preservation (Curtis, 1993). Described above, Tombstone, Arizona, is home to several buildings on this register. This brand is more inclusive than the NHLs, including many broader categories of historic places, although all NHLs are also on the National Register. At the end of 2023, there were more than 98,000 properties on the register (National Park Service, 2023b). This inventory aims to highlight as many historic sites and places as possible to assist in their preservation, and almost every US county is represented on the list. Individual property owners can apply to have their sites listed, but they are responsible for paying for their own commemorative plaques or markers. They may apply for funding to help maintain or repair an historic property, but the primary responsibility is that of the owners. Despite including so many properties and sites, including many small-scale localities and structures, the National Register brand is also believed to have significant tourism value for the communities that host them (Hargrove, 2002; Rypkema, 2002).

Properties identified by the United States' National Trust for Historic Preservation (NTHP) are also heritage branded. The NTHP is a non-profit, privately funded organization that aims to preserve and protect the nation's historic buildings, neighborhoods, and other heritage assets through advocacy and specific action programs. The organization frequently partners with the National Park Service and other agencies to raise funds, enact protective legislation, and draw attention to America's historic resources. Its goals include saving America's historic sites, telling the full American story, building stronger communities, and investing in preservation's future (National Trust for Historic Preservation, 2023). The organization today operates 27 sites open for public visits. As part of its mandate, NTHP has had, and continues to maintain, several important programs and emphasis areas that deal with African American heritage, women making history, sacred places, Main Street America, endangered places, historic hotels, and national treasures.

Recovery Branding

A different perspective on heritage and branding is the notion of using heritage to help places in crisis reimage or rebrand themselves as viable, safe destinations. This is a common recovery effort in places that have suffered natural disasters, political crises, or other human-induced catastrophes (Avraham, 2019; Seraphin et al., 2018). In the post-COVID environment, several destinations that traditionally have not had a strong heritage tourism sector, have now turned to heritage as a coping and recovery mechanism. Dubai is a prime example of this phenomenon, as it endeavored to supplement its 'luxury' brand by emphasizing the emirate's heritage potential since 2021 (Haq et al., 2023).

Other countries have turned to heritage tourism to revive their economies after long-term conflicts. Nicaragua, Peru, Haiti and Croatia, for instance, have all used heritage-based tourism in recent years to turn tourists' attention away from the countries' histories of violence, civil wars, and revolutionary conflicts. In these four countries, heritage branding in tourism is believed to educe an image of a safer and healthier destination and can return tourism as a viable economic activity in post-conflict environments (Babb, 2020; Seraphin et al., 2018; Wise & Mulec, 2012).

Similarly, Ghaderi et al. (2012) showed how heritage became a means of recovering from the 2011 Indian Ocean tsunami, which destroyed many of the beaches that put Penang, Malaysia, on the tourist map. This disaster provided an opportunity for officials and stakeholders to rethink, reflect, and reassess their tourism sector, with a greater branding effort devoted to the island's cultural heritage.

Although deflecting from negative publicity and recovering from formerly dangerous conditions through rebranding post-disaster and post-conflict regions with heritage tourism labels is common, another outcome is growing in significance as market demand for extraordinary and unusual experiences has changed in recent decades. Although dark heritage (i.e., sites of disaster, death, and human suffering) has long captivated the attention of tourists, this form of heritage tourism is now commonplace and has received widespread promotional and research attention (Hartmann, 2014; Light, 2017; Stone, 2023). The 'dark tourism' brand is something many post-disaster or post-conflict destinations have turned to in an effort to resurrect their damaged tourism industries. Just like branding efforts in general, this sort of action can be either intentional, organic, or a combination of the two (Rindrasih & Witte, 2021).

Following a disaster or major catastrophe, it is natural for onlookers to appear, gawking at the damage and voyeuristically gazing upon the suffering of others. This often evolves into dark tourism, which brings its own negative connotations and naysayers, albeit with some positive economic impacts. DMOs and other stakeholders sometimes grasp the potential of this sort of shadowed tourism and begin to promote their localities as dark destinations. This sort of morbid tourism has the potential to help a destination recover and rebrand itself based on the misfortune of those who died or were injured (Neef & Grayman, 2018).

'Disaster tourism' in Jina Qiang Village, China, following the devastating 2008 earthquake that killed 30,000 people, or 10% of the region's population, was seized upon by local authorities and helped to bring the Qiang ethnic minority out of obscurity into mainstream tourism. The tourism stimulated by the disaster has helped restore the impoverished community and even helped resurrect some Qiang cultural traditions that had begun to waver (Zhang, 2012). A similar situation emerged in Nepal after the devastating earthquakes there in 2015. Disaster-based dark tourism has been lauded as a mechanism for disaster recovery and has spread to include other types of tourism to accompany it (Min et al., 2020; Nepal & Devkota, 2018). While disaster tourism is hardly desirable, it is an example of places being willing to capitalize in their misfortunes to create a unique heritage brand that can help raise the destination out of its disaster conditions and increase a level of global visibility that might not have been as known before the disaster occurred. In the words of Rindrasih and Witte (2021, p. 136), "... the earthquake [in Yogyakarta, Indonesia] has refreshed the cultural landscape, and ... new tourism packages [use] the story of heritage in combination with the earthquake even [leading] to new economic opportunities".

Even tourism based on human-caused calamities (e.g., terrorism, war, and other crises) can effectively help a destination rise above its calamitous state. Northern Ireland's longtime conflict ('the troubles') is now an important part of the country's heritage narrative (Reddy et al., 2020), and the narco heritage of Medellin, Colombia, and sites there associated with drug king Pablo Escobar, are an important part of the tourism landscape in that South American city (Naef, 2018).

A final perspective on recovery branding is the notion of declining industrial communities that have turned to tourism as a means of replacing what has been lost in their economic landscapes. For instance, many old mining or timber towns facing severe socio-economic declines have turned to tourism as a means of staying afloat or reinvigorating their economic health. The tourism associated with the majority of these towns and villages is heavy on heritage and retail, with most of them being classified as 'tourist shopping villages' (TSVs) (Getz, 1993; Murphy et al., 2011). Characteristics in common among almost all TSVs include some sort of industrial or agricultural past that has now diminished, causing the community to reorient itself into a day-trip or weekend visitor destination, typically near larger cities and good access routes. By rebranding themselves as shopping villages with strong heritage undertones, these TSVs have seen considerable success throughout the USA, Australia, Canada, and in many other countries with post-Fordist declining economies (Timothy, 2021).

Conclusion

Heritage tourism has come to represent the tourism industry at large in many parts of the world. Some countries, regions, and cities are nearly synonymous with their heritage milieu and are sought primarily for their cultural resources. This process often takes place in an organic manner as historic locales of critical acclaim appeal to outside observers through time, eventually becoming significant tourist destinations. Other destinations have come to realize the value of heritage and have actively selected this asset as a branding mechanism to enhance their economic growth and build national pride. What we inherit from the past, utilize and value today, and hope to pass on to future generations—heritage—is a crucial part of every place because of the rootedness it provides and the sense of social solidarity that it can encourage. These characteristics and others are essential to the successful development of cultural heritage-based tourism.

Places frequently hone their branding efforts in on one or a few historic sites or cultural features, often ignoring other important elements of the heritagescape that may be equally important. For example, India has long relied on the image of the Taj Mahal for much of its brand identity, yet the country has considerable potential to enjoin many of its other heritage features, such as places and events associated with Mahatma Gandhi, in its global marketing efforts (Chhabra, 2017). Nonetheless, sharing the global image of India with other valuable historic features is easier said than done. In many countries, branding some historic sites while not branding others may create imbalances and inequities that are difficult to overcome. National labels and international brands can impart an unfair advantage to places that already have a well-developed tourism sector, thus empowering some communities while disempowering others through heritage commemoration and branding.

There are hundreds of official mechanisms, or registers, at the global scale and at most national levels that identify heritage districts, sites, and objects as important, authentic, and worthy of protection. Many of these same brands function simultaneously, but often unintentionally, as tourism brands that garner the attention of government agencies, DMOs, and tourists. Such brands are believed to support many of the principles that underlie official heritage designations, including sustainability, historical integrity and authenticity, preservability, societal ideals, and economic value. Because of this, they also frequently exude considerable tourism appeal. Heritage branding also has the potential to help destinations rebound from crises and reorient themselves in a way that expands their existing tourism offer and provides alternative place images to the ravages of natural or human-induced disasters.

Although much has been written about the potential brand value of global and national heritage designations, more work is needed to understand more clearly different senses of place with regard to heritage, different cultural contexts and backgrounds in how heritage brands are esteemed, the effects of heritage valorization on tourism, and the added economic value of different heritage brands.

References

Adie, B. A. (2017). Franchising our heritage: The UNESCO World Heritage brand. *Tourism Management Perspectives*, *24*, 48–53.

Adie, B. A., & Hall, C. M. (2017). Who visits World Heritage? A comparative analysis of three cultural sites. *Journal of Heritage Tourism*, *12*(1), 67–80.

Amer, M., Ginzarly, M., & Renzi, M. F. (2023). Civita di Bagnoregio, Italy: Towards a people-centred heritage branding approach. *Journal of Heritage Tourism*, *18*(4), 483–503.

Avraham, E. (2019). Image repair strategies adopted by MENA national tourism boards during and following crises. In D. J. Timothy (Ed.), *Routledge handbook on tourism in the Middle East and North Africa* (pp. 241–256). Routledge.

Babb, F. E. (2020). Peru and Nicaragua: Tourism development in post-conflict eras. In C. Monterrubio, K. Antriotis, & D. Stylidis (Eds.), *Tourism planning and development in Latin America* (pp. 156–172). CABI.

Bhakuni, K., & Kapoor, S. (2011). The marvel of Taj: A competitive in brand positioning in tourism. *Business Analyst*, *40*(1), 45–57.

Boggs, J. D. (2019). *The American West on film*. Bloomsbury.

Chhabra, D. (2017). Heritage branding of India: A Gandhi tourism perspective. *Annals of Tourism Research*, *62*, 110–112.

City of Tombstone (2012). *City of tombstone state route 80 alternate route PARA study*. Arizona Department of Transportation.

Council of Europe (2023). Cultural routes. Online: https://www.coe.int/en/web/cultural-routes/all-cultural-routes

Cullingworth, J. B. (1992). Historic preservation in the US: From landmarks to planning perspectives. *Planning Perspective*, *7*(1), 65–79.

Curtis, M. E. (1993). A heritage tourism tool: National register of historic places. In G. Vander Stoep (Ed.), *Proceedings of the 1992 northeastern recreation research symposium* (pp. 73–74). US Forest Service, Northeastern Forest Experiment Station.

De Simone, E., Canale, R. R., & Di Maio, A. (2019). Do UNESCO World Heritage Sites influence international tourist arrivals? Evidence from Italian provincial data. *Social Indicators Research*, *146*, 345–359.

El Daly, O. (2016). What do tourists learn from Egypt? In S. MacDonald, & M. Rice (Eds.), *Consuming ancient Egypt* (pp. 139–150). Routledge.

European Commission (2023). European capitals of culture. Online: https://culture.ec.europa.eu/policies/culture-in-cities-and-regions/european-capitals-of-culture

Flores-Romero, M. B., Perez-Romero, M. E., Alvarez-Garcia, J., & del Río-Rama, M. D. L. C. (2021). Bibliometric mapping of research on magic towns of Mexico. *Land*, *10*(8), 852.

Foroudi, P., Mauri, C., Dennis, C., & Melewar, T. C. (Eds.). (2020). *Place branding: Connecting tourist experiences to places*. Routledge.

Frank, K., & Petersen, P. (2002). *Historic preservation in the USA*. Springer.

Getz, D. (1993). Tourist shopping villages: Development and planning strategies. *Tourism Management*, *14*(1), 15–26.

Ghaderi, Z., Som, A. P. M., & Henderson, J. C. (2012). Tourism crises and island destinations: Experiences in Penang, Malaysia. *Tourism Management Perspectives*, 2, 79–84.

Hall, C. M., & Piggin, R. (2002). Tourism business knowledge of World Heritage Sites: A New Zealand case study. *International Journal of Tourism Research*, *4*(5), 401–411.

Hamid, S. A., Abdullah, Q. D. L., & Lee, T. M. (2018). Destination image of UNESCO World Heritage Sites in Malaysia. *Global Business & Management Research*, *10*(2), 22–30.

Haq, F., Seraphim, J., & Medhekar, A. (2021). Branding heritage tourism in Dubai: A qualitative study. *Advances in Hospitality and Tourism Research*, *9*(2), 243–267.

Haq, F., Yasin, N., & Nair, G. (2023). Envisioning the future of heritage tourism in the creative industries in Dubai: An exploratory study of post COVID-19 strategies for sustainable recovery. *Heritage*, *6*(6), 4557–4572.

Hargrove, C. M. (2002). Heritage tourism. *Cultural Resource Management*, *25*(1), 10–11.

Hartmann, R. (2014). Dark tourism, thanatourism, and dissonance in heritage tourism management: New directions in contemporary tourism research. *Journal of Heritage Tourism*, *9*(2), 166–182.

Hernández, M. M. (2019). Sustainable tourism or selling places: Effects of the UNESCO World Heritage Site nomination in Valparaíso and Hoi An. *International Journal of Heritage and Museum Studies*, *1*(1), 138–147.

Herstein, R., & Berger, R. (2013). Hosting the Olympics: A city's make-or-break impression. *Journal of Business Strategy*, *34*(5), 54–59.

Hussin, H. (2018). Branding Malaysia and re-positioning cultural heritage in tourism development. *JATI-Journal of Southeast Asian Studies*, *2018*, 75–92.

Iaţu, C., Ibănescu, B. C., Stoleriu, O. M., & Munteanu, A. (2018). The WHS designation—A factor of sustainable tourism growth for Romanian rural areas? *Sustainability*, *10*(3), 626.

Jiménez Madariaga, C. D., & Seño Asencio, F. (2019). "We are brand." Tourism and UNESCO brand in the intangible cultural heritage. *PASOS: Revista de Turismo y Patrimonio Cultural*, *17*(6), 1127–1141.

Jimura, T. (2016). World Heritage Site management: A case study of sacred sites and pilgrimage routes in the Kii Mountain Range, Japan. *Journal of Heritage Tourism*, *11*(4), 382–394.

Jimura, T. (2022). *Cultural heritage and tourism in Japan*. Routledge.

Journey Mexico (2023). Pueblos Magicos: A guide to Mexico's magic towns. Online: https://www.journeymexico.com/blog/pueblos-magicos-mexico-magic-towns

Kim, E. G., Chhabra, D., & Timothy, D. J. (2022). Towards a creative MICE tourism destination branding model: Integrating heritage tourism in New Orleans, USA. *Sustainability*, *14*(24), 16411.

King, L. M., & Halpenny, E. A. (2014). Communicating the World Heritage brand: Visitor awareness of UNESCO's World Heritage symbol and the implications for sites, stakeholders and sustainable management. *Journal of Sustainable Tourism*, *22*(5), 768–786.

Light, D. (2017). Progress in dark tourism and thanatourism research: An uneasy relationship with heritage tourism. *Tourism Management*, *61*, 275–301.

Lin, Y. X., Lin, B. S., Chen, M. H., & Su, C. H. (2020). 5A tourist attractions and China's regional tourism growth. *Asia Pacific Journal of Tourism Research*, *25*(5), 524–540.

Martens, T., & Timothy, D. J. (2006). Burying tombstone: Fading authenticity in a Wild West tourist destination. Paper presented at the American Association of Geographers Annual Conference, Chicago, Illinois, March 7–10.

Maskey, V., Brown, C., & Lin, G. (2009). Assessing factors associated with listing a historic resource in the National Register of Historic Places. *Economic Development Quarterly*, *23*(4), 342–350.

McCormack, K. L. (2016). *Imagining tombstone: The town too tough to die*. University Press of Kansas.

Meskell, L. (2018). *A future in ruins: UNESCO, World Heritage, and the dream of peace*. Oxford University Press.

Meskell, L., Liuzza, C., Bertacchini, E., & Saccone, D. (2015). Multilateralism and UNESCO World Heritage: Decision-making, states parties and political processes. *International Journal of Heritage Studies*, *21*(5), 423–440.

Min, J., Kc, B., Kim, S., & Lee, J. (2020). The impact of disasters on a heritage tourist destination: A case study of Nepal earthquakes. *Sustainability*, *12*(15), 6115.

Ministry of Tourism. (2023). Ecuador and its 21 Magical Towns with so much to discover. Online: https://ecuador.travel/ecuador-y-sus-21-pueblos-magicos-con-tanto-por-descubrir/

Morgan, N., Pritchard, A., & Pride, R. (2004). *Destination branding: Creating the unique destination proposition* (2nd ed.). Elsevier.

Mota, M., Guerreiro, M., & Pinto, P. (2017). The world heritage brand and tourism: An approach to the historic centre of São Luís, Brazil—Heritage tourism marketing. In C. Henriques, M. C. Moreira, & P. Bittencourt César (Eds.), *Tourism and history: World heritage—Case studies of Ibero-American space* (pp. 287–311). Interdisciplinary Centre of Social Sciences.

Murphy, L., Moscardo, G., Benckendorff, P., & Pearce, P. (2011). Evaluating tourist satisfaction with the retail experience in a typical tourist shopping village. *Journal of Retailing and Consumer Services*, *18*(4), 302–310.

Mustafa, M. H. (2021). Cultural heritage: A tourism product of Egypt under risk. *Journal of Environmental Management and Tourism*, *12*(1), 243–280.

Naef, P. (2018). Touring the "comuna": Memory and transformation in Medellin, Colombia. *Journal of Tourism and Cultural Change*, *16*(2), 173–190.

National Park Service (2023a). National Historic Landmarks Program. Online: https://www.nps.gov/orgs/1582/index.htm

National Park Service (2023b). National Register of Historic Places. Online: https://www.nps.gov/subjects/nationalregister/index.htm

National Trust for Historic Preservation (2023). Our work. Online: https://savingplaces.org/

Nazir, F., Caldeira, A. M., & Seabra, C. (2023). Heritage tourism and terrorism: Media coverage of the destruction and rebuilding of Jahanabad seated Buddha in Pakistan. *Journal of Heritage Tourism*, *18*(4), 504–530.

Neef, A., & Grayman, J. H. (2018). Conceptualising the tourism–disaster–conflict nexus. In A. Neef, & J. H. Grayman (Eds.), *The tourism–disaster–conflict nexus* (pp. 1–31). Emerald.

Nepal, S. K., & Devkota, B. (2018). Post-disaster recovery, tourism, and heritage conservation: Insights from the impacts of Nepal's 2015 earthquake in Kathmandu Valley. In J. Saarinen, & A. M. Gill (Eds.), *Resilient destinations and tourism: Governance strategies in the transition towards sustainability in tourism* (pp. 106–120). Routledge.

Olivé, C. G. A., & Torres-González, A. (2023). Social networks as a positioning strategy of a magic town. *Journal of Tourism and Heritage Research*, *6*(4), 96–114.

Patel, K. K. (Ed.) (2013). *The cultural politics of Europe: European capitals of culture and European Union since the 1980s*. Routledge.

Reddy, M. V., Boyd, S. W., & Nica, M. (2020). Towards a post-conflict tourism recovery framework. *Annals of Tourism Research*, *84*, 102940.

Rindrasih, E., & Witte, P. (2021). Reinventing the post-disaster cultural landscape of heritage tourism in Kotagede, Yogyakarta, Indonesia. *Journal of Heritage Tourism*, *16*(2), 136–150.

Ryan, C., Huimin, G., & Meng, F. (2009). Destination planning in China. In C. Ryan, & G. Huimin (Eds.), *Tourism in China: Destination, cultures and communities* (pp. 11–37). Routledge.

Ryan, J. (2015). Intangible cultural heritage: The new frontier of destination branding. In K. Kubacki (Ed.), *Ideas in marketing: Finding the new and polishing the old* (pp. 388–390). Springer.

Ryan, J., & Silvanto, S. (2009). The World Heritage List: The making and management of a brand. *Place Branding and Public Diplomacy*, *5*, 290–300.

Ryan, J., & Silvanto, S. (2010). World Heritage Sites: The purposes and politics of destination branding. *Journal of Travel & Tourism Marketing*, *27*(5), 533–545.

Rypkema, D. D. (2002). The (economic) value of National Register listing. *Cultural Resource Management*, *25*(1), 6–7.

Saeedi, H., & Heidarzadeh Hanzaee, K. (2018). The effects of heritage image on destination branding: An Iranian perspective. *Journal of Heritage Tourism*, *13*(2), 152–166.

Santamarina, B. (2023). The global competition of the intangible. UNESCO as a producer of heritage brands. *Heritage & Society*, *16*(3), 251–270.

Secretaría de Turismo (2016). Programa Pueblos Mágicos. Online: https://www.gob.mx/sectur/acciones-y-programas/programa-pueblos-magicos

Seraphin, H., Yallop, A. C., Capatîna, A., & Gowreesunkar, V. G. (2018). Heritage in tourism organisations' branding strategy: The case of a post-colonial, post-conflict and post-disaster destination. *International Journal of Culture, Tourism and Hospitality Research*, *12*(1), 89–105.

Severo, M. (2017). European cultural routes: Building a multi-actor approach. *Museum International*, *69*(1–2), 136–145.

Shabani, B., Tucker, H., & Nazifi, A. (2021). UNESCO's World Heritage Sites: The interplay between international and local branding for the Gonbad-e Qābus Brick Tower, Iran. In C. M. Hall, & S. Seyfi (Eds.), *Cultural and heritage tourism in the Middle East and North Africa: Complexities, management and practices* (pp. 187–201). Routledge.

Stone, P. (2023). Dark tourism and 'painful pasts' in Africa: Concepts, contexts, and challenges. In D. J. Timothy (Ed.), *Cultural heritage and tourism in Africa* (pp. 217–238). Routledge.

Timothy, D. J. (1997). Tourism and the personal heritage experience. *Annals of Tourism Research*, *34*(3), 751–754.

Timothy, D. J. (2018). Producing and consuming heritage tourism: Recent trends. In S. Gmelch, & A. Kaul (Eds.), *Tourists and tourism: A reader* (3rd ed., pp. 167–178). Waveland Press.

Timothy, D. J. (2021). *Cultural heritage and tourism: An introduction* (2nd ed.). Channel View Publications.

Timothy, D. J., & Boyd, S. W. (2003). *Heritage tourism*. Prentice Hall.

Timothy, D. J., & Boyd, S. W. (2006). Heritage tourism in the 21st century: Valued traditions and new perspectives. *Journal of Heritage Tourism*, *1*(1), 1–16.

Timothy, D. J., & Boyd, S. W. (2015). *Tourism and trails: Cultural, ecological and management issues*. Channel View Publications.

Timothy, D. J., & Tahan, L. G. (2020). Archaeology and tourism: Consuming, managing and protecting the human past. In D. J. Timothy, & L. G. Tahan (Eds.), *Archaeology and tourism: Touring the past* (pp. 1–25). Channel View Publications.

Tsaur, S. H., Liao, Y. L., & Tsai, C. F. (2020). Analyzing the important implications of tourism marketing slogans and logos in Asia Pacific nations. *Asia Pacific Journal of Tourism Research*, *25*(4), 355–368.

Wang, J., Huang, X., Gong, Z., & Cao, K. (2020). Dynamic assessment of tourism carrying capacity and its impacts on tourism economic growth in urban tourism destinations in China. *Journal of Destination Marketing & Management*, *15*, 100383.

Wise, N. A., & Mulec, I. (2012). Headlining Dubrovnik's tourism image: Transitioning representations/narratives of war and heritage preservation 1991–2010. *Tourism Recreation Research*, *37*(1), 57–69.

Zhang, L., & Zhao, S. X. (2009). City branding and the Olympic effect: A case study of Beijing. *Cities*, *26*(5), 245–254.

Zhang, Q. (2012). Phoenix out of the ashes: Convergence of disaster and heritage tourism in Jina Qiang Village, Sichuan Province, China. *Practicing Anthropology*, *34*(3), 24–28.

9

REBRANDING OR RETRO-BRANDING OF HERITAGE DESTINATIONS

Orhan Can Yılmazdoğan and Cemile Ece

Introduction

Heritage tourism is the practice of touring historical and cultural landmarks. Such tourism-related activities are typically done to discover historical and culturally significant locations, to take in the customs, handicrafts, architecture and heritage of the area (Laws, 2007). Heritage tourism helps the local economy, but it also promotes the preservation of cultural and historical assets and the growth of a sustainable tourism industry. Tourists must preserve the maintenance of cultural heritage, the environment and the local economy, while they are visiting in order for tourism operations to be viable. Heritage tourism destinations need to be branded in order to attract more tourists (Gülmez et al., 2012). Destinations that develop a branding strategy can use different strategies and methods. Retro-branding and rebranding are two important marketing strategies that can be used in heritage destination marketing to attract more tourists and stimulate economic growth. Retro-branding involves reviving a brand or product from a previous era, while rebranding involves changing the brand image and identity of a destination. This chapter covers the use of retro-branding and rebranding in heritage destination marketing strategies and examples of good practices are also included to better examine their potential benefits and challenges.

Heritage Destinations

Heritage tourism is a growing industry that aims to promote and protect cultural and historical heritage landmarks (Huh et al., 2006). Cultural and historical heritage sites are preserved and handed on to future generations

DOI: 10.4324/9781003346197-9

with the assistance of this type of tourism, which also supports local communities economically. The potential to increase local economies and generate employment is one of the major advantages of heritage tourism. Local communities can assist tourists who want to learn about local history and culture by promoting heritage sites. This promotes new employment opportunities in tourism-related sectors. The economic well-being of nearby communities improves with the increase of new employment opportunities, and new service areas can be established in the heritage region (Hargrove, 2002). Because of this, heritage tourism promotes essential initiatives that ensure local development. Local communities educate the community regarding the importance of protecting historical landmarks and cultural traditions by promoting cultural and historical heritage sites.

However, heritage tourism can cause negative effects if not managed carefully. Overcrowding and overuse of heritage sites can cause environmental damage, degradation of cultural monuments and other adverse effects on the local community (Andriotis et al., 2021). Therefore, it is necessary to develop sustainable tourism practices that balance the economic benefits of tourism. Heritage tourism, especially in destination management, should adopt sustainable tourism management. By sustainable destination management, heritage tourism destinations can protect the necessary development in environmental, economic and social terms and create opportunities for local communities by providing local development, while preserving the cultural identity.

Tourism mobility and policies developed in heritage destinations ensure the preservation of cultural and historical sites. Especially within national policies, numerous measures are observed to be developed for the preservation of cultural and historical values. Through these laws, efforts are made to ensure sustainability by protecting both tangible and intangible cultural elements in heritage destinations and transferring them to future generations (Giudici et al., 2013). The accessibility of heritage destinations and the high quality of tourism infrastructure can attract more tourists to the region, contributing to improved transportation, accommodation and dining services as well as facilitating economic benefits for local businesses. The most significant regional contribution of heritage destinations is their support for local communities and environmental regulations, which, in turn, foster the development of sustainable tourism practices. Thanks to tourism activities carried out in heritage destinations, local communities participate, enabling them to benefit from tourism and share their own cultures and histories with visitors (Su et al., 2020).

In order to realize the stated benefits of heritage destinations, effective strategies need to be developed by tourism managers and local authorities to preserve and promote heritage sites, and these strategies should collectively contribute to the region's marketing (Rahman et al., 2021). Marketing encompasses all necessary strategies for promoting heritage destinations, aiding them in reaching a wider audience and differentiating from competitors. Through marketing, heritage destinations showcase their unique features and tourist attractions to potential visitors, thereby creating brand value.

Destination branding is crucial for them to gain a share in the tourism market and to promote local development. Effective destination branding in heritage sites requires highlighting the unique features of the destination and creating a distinctive brand image. This way, by emphasizing a prominent aspect during the branding process, destinations can achieve a competitive advantage in the tourism market and attract visitors to the region throughout the year. Developing a strong brand identity is essential for destination branding, allowing destinations to differentiate themselves from competitors and create a distinct image in the minds of visitors (Lin et al., 2020). Effective brand awareness also aids in building a customer base for destinations.

Heritage destinations can create a unique and memorable experience for visitors, fostering positive associations and increasing the likelihood of repeat visits. Moreover, establishing brand awareness can have positive effects on marketing heritage destinations and enhancing customer satisfaction (Paddison & Biggins, 2017). For destinations to compete with other rivals and enhance tourism activities, possessing a well-defined branding strategy is essential.

Rebranding vs. Retro-branding: Pros and Cons

Branding aims to distinguish and promote a product, service or organization by using its name, logo, design and other features. Branding increases the perceived value of the product or service and is a tool that facilitates consumers' knowledge and experience of the product or service (Murphy, 1988). Successful branding enables a brand to create a unique identity and value proposition and helps convey the right messages to the target audience. The fact that branding has become important recently with the effect of social media has also revealed different approaches. When the activities and marketing studies in the field of tourism are examined, the concepts of retro-branding and rebranding come to the fore.

Retro-branding is defined as a brand identity or marketing strategy that refers to a brand's past or is redesigned in a nostalgic style (Brown et al., 2003). Retro-branding often includes designs and marketing strategies using unique or iconic elements from the brand's history. This type of brand identity design attracts consumers who yearn for a period or style in the brand's history. Retro-branding also conveys a message to consumers about the reliability and quality of the brand, referring to the brand's long history or longevity (Chen, 2022). Retro-branding, which is frequently used in the promotion of tourism products and destinations, is a very suitable marketing strategy for heritage tourism. Heritage tourism attracts tourists who want to visit places of cultural or historical value. These tourists are usually people who are interested in the past, admire the cultural riches and want to explore the local history. For this reason, heritage tourism areas benefit greatly from the retro-branding strategy. Heritage tourism sites use retro-branding strategy by emphasizing elements related to local culture and history, such as historical buildings, local handicrafts, traditional food, local clothing or museums (Seraphin et al., 2018). These elements can be presented in a way that reflects the history of the brand, respects the local culture and tradition and provides tourists with a nostalgic experience (Balmer, 2011). However, while using the retro-branding strategy, the modern and innovative aspects of heritage tourism areas should not be ignored. While the retro-branding strategy is a tool used to attract tourists, it should also reflect the quality and timeliness of the brand.

Another marketing tool used in the field of branding is rebranding. Rebranding means changing a brand's existing identity or marketing strategy. Rebranding is a preferred branding management for many reasons, such as brand renewal, restructuring or redefining a brand in the process of change. In the rebranding process, the brand's current identity and marketing strategy are examined and changes are made to create an identity and strategy that are more suitable for the brand's new target audience or goals (Goi & Goi, 2011). Changes made on elements such as the name, logo, colors, slogan or marketing message of the brand are shown as examples. Rebranding is a concept that has started to be used in the field of tourism as well. Heritage tourism areas, in particular, use rebranding strategy to attract the attention of tourists by designing a new brand identity and renewing their brands. By having a new brand image, it contributes to the growth of heritage tourism (Mutana & Zinyemba, 2013).

Although rebranding and retro-branding are different brand strategies, both are used to change or revitalize the brand's existing identity. However, both branding strategies have pros and cons (see Table 9.1).

Rebranding and retro-branding are different brand strategies used to change or revitalize the existing identity of the brand. For this reason, the branding strategies used in heritage destinations have different approaches and principles.

Strategies for Rebranding or Retro-branding of Heritage Destinations

Heritage destinations are areas that provide its guests with a distinctive tourism experience while showcasing the nostalgic, historical and cultural qualities of a location. However, due to changing visitor preferences and the need to remain relevant, rebranding or retro-branding of heritage destinations has become a common practice in the tourism industry. Rebranding

TABLE 9.1 Pros and Cons of Rebranding and Retro-branding

	Pros	*Cons*
Retro-Branding	• is a marketing strategy based on nostalgia, it can increase the customer loyalty of the brand. • Emphasize the roots of the brand by respecting the history and background of the brand. • can attract old customers again.	• may not be effective at attracting new customers. • The brand runs the risk of losing its old customers. • The process of establishing a historical identity may conflict with modern brand values. • appeals only to a specific target audience, so it is not suitable for reaching large audiences.
Rebranding	• helps to renew and modernize the brand. • can be used to reach a new target audience. • is an effective strategy to stand out among competitors. • can help reinvigorate the brand's reputation. • can help fix an old or negative image of the brand.	• The brand runs the risk of losing existing customers. • The process of creating a new brand identity is costly. • The process of creating a new brand identity is time consuming. • Taking wrong steps in the process of creating a new brand identity can negatively affect the brand's reputation.

and retro-branding are strategies used to revitalize heritage destinations. Rebranding involves updating the visual identity, marketing strategies and visitor experiences of a place to appeal to modern travelers. The goal of rebranding is to create a unique identity for a destination that reflects its cultural and historical significance while offering modern amenities and experiences to visitors (Kim et al., 2018). However, retro-branding involves reviving the past and portraying it in a modern context. Retro-branding is ideal for heritage destinations that have lost their cultural and historical significance due to modernization and its aim is to restore the historical and cultural significance of a place while offering modern amenities and experiences to visitors. Thus, while associating a product with the relevant historical period, it strengthens the relations of heritage brands with the period in which they existed in the past, by ensuring that the product preserves its classic features (Chen, 2022). The primary difference between rebranding and retro-branding is their focus. While rebranding focuses on modernizing a destination to appeal to modern travelers, retro-branding focuses on restoring the cultural and historical significance of a place. However, the aim of both strategies is to achieve a balance between preserving the cultural and historical aspects of a destination and modernizing it (Kumar et al., 2019).

A lot of research has been carried out on rebranding and retro-branding through heritage destinations. According to Li et al. (2013) in one of these studies, heritage destinations that want to brand should be able to establish an emotional connection with visitors while exhibiting their historical and cultural aspects. In another study conducted in the Wulingyuan region of China, Wang et al. (2016) stated that more tourists and tourism income can be possible through rebranding, and this can be achieved through the preservation of cultural and historical features while providing modern facilities to visitors. Similarly, Kockelman and Lohr (2019) stated the importance of redesigning a destination in the context of heritage tourism, redesigning it creatively, benefiting from rebranding and retro-branding in this, and they stated that this should be done in a way that the destination can create a unique identity while reflecting the cultural and historical importance. Gursoy and Mcleod (2019) similarly emphasized that rebranding and retro-branding will contribute to creating a unique identity for these destinations while attracting visitors to heritage tourism destinations. Based on the recommendations of this and similar studies and current practices, we can list the following strategies for rebranding and retro-branding:

Rebranding Strategies

Rebranding of heritage destinations involves changing the visual identity, marketing strategies and visitor experiences to appeal to modern travelers. As the researchers emphasized in their studies, rebranding provides various advantages to destinations, especially in terms of cultural heritage tourism, while at the same time mediating the preservation of cultural and historical attractions. Some of the rebranding strategies used for this purpose are as follows:

Creating a Unique Identity

One of the important strategies that researchers have emphasized in their research is the creation of a unique identity for a destination through rebranding. Doing this is extremely important in attracting visitors to the destination and gaining competitive advantage. While doing this, the aspects that make the destination privileged should be emphasized and cultural and historical richness should be reflected. It has been observed that destinations that can achieve this increase their tourism revenues by attracting more visitors. At this point, while emphasizing the cultural and historical aspects specific to the destination, it is important to preserve originality and avoid imitation.

Preserving Cultural and Historical Significance

While carrying out activities related to rebranding, it is extremely important to choose appropriate strategies to reflect the cultural and historical significance of the destination. While taking steps toward modernization, it should be done in accordance with its historical and cultural importance. The balance here is extremely sensitive and important, while scientific studies underline this point, they state that overemphasis on modernization may cause loss of cultural and historical significance. In this respect, it is necessary to adopt an approach based on balancing historical and cultural significance with modern opportunities in rebranding strategies, and it also contributes to originality.

Collaboration with Stakeholders

Collaboration with stakeholders stands out as an important requirement for more harmonious, sustainable and successful remarketing strategies. Researches also indicate that rebranding should include stakeholders in the destination and should be planned holistically. To this end, destinations can develop partnerships with local restaurants, hotels, non-governmental organizations and other tourist destinations to provide a more attractive and comprehensive tourist experience. Achieving this will not only support a holistic destination management, but also help rebrand strategies to be adopted by stakeholders and welcomed and supported by local communities.

Revamping the Visitor Experience and Embracing Technology

Combining the traditional with the modern in a balanced way and presenting it correctly is an important step in attracting modern visitors and enriching the visitor experience. In their rebranding strategies, heritage destinations should pay attention to presenting modern opportunities by making optimum use of technology. For example, the former glorious periods of heritage destinations and the structure of that period can be reflected by supporting them with technological opportunities such as virtual reality and augmented reality applications, and comparisons can be made with today. Technology needs to be incorporated into the rebranding process to increase the visitor experience and make it immersive and memorable. Augmented and virtual reality applications, smart and wearable technologies, interactive spaces and displays, robots and artificial intelligence are constantly offering new and exciting alternatives to engage visitors and enrich their experience. Destinations that evaluate them and include them in the rebranding process will stand out in competition and create a strong destination image. Another important requirement is to present the features of these destinations to potential tourists with a successful social media marketing. Micro and macro influencers are very helpful in communicating the rebranding processes of destinations to large audiences.

Sustainability

Another important strategy is for destinations to include sustainable and environmentally friendly practices in their rebranding processes. While destinations draw attention to their historical and cultural significance and combine it with modernization in a balanced way, they can brand themselves by sharing what steps they have taken regarding sustainability in response to increasing environmental concerns. Steps taken/planned to be taken regarding sustainability may include promoting sustainable tourism practices such as using renewable energy sources, reducing waste and carbon footprint and sustainable and green procurement.

Personalization

Personalization is becoming more and more important in all sectors and especially in tourism. While visitors care that their visits are unique and memorable, they also question how much this appeals to their personal characteristics. Based on these current expectations, heritage destinations can offer tours and experiences customized to their visitors' expectations through the data they collect from and analyze their visitors.

Storytelling

One of the important tools that heritage destinations can use in their rebranding processes is storytelling. Visitors care about the people, events, places and interactions involved in the destination's history from past to present. Engaging visitors with stories about these components can create a deeper understanding and appreciation of the destination's heritage, history and culture. This can happen through the use of both traditional methods and digital technologies, guided tours, interactive exhibits and displays.

Retro-branding Strategies

Retro-branding involves portraying the past for a heritage destination while portraying its nostalgic features in a modern context. Especially against the risk of losing its cultural and historical importance with the negative effect of modernization, retro-branding is ideal for heritage destinations. In this regard, some retro-branding strategies that can be used for heritage destinations are included:

Restoring the Original Architecture

One of the most important retro marketing strategies is restoring cultural heritage items in a destination according to their original architecture. For destinations that are moving away from their original and cultural texture with the effect of modernization, restorations and reconstructions made according to the original architecture can help preserve and reflect the culture and history of the destination in an authentic way.

Offering Traditional Experiences

In retro-branding, areas such as local cuisine and handicrafts can be used to present traditional experiences and mediate the revival of the past in the experiences of visitors. For example, while Registan Square in Samarkand, Uzbekistan, was renovated in accordance with its nostalgic original, it was also supported with places where local cuisine and handicrafts can be presented. Offering traditional and authentic experiences can help preserve the cultural and historical significance of cultural heritage destinations and pass them on to visitors and future people.

Collaboration with Local Communities

Collaboration with local communities and tourism stakeholders is an important requirement in retro-branding as in rebranding. Holistic practices with local communities and stakeholders can help develop a successful marketing campaign while providing insight into the cultural and historical significance of the place. These collaborations can contribute to the adoption of retro-branding strategies and increase my support for it.

Developing Themed Events and Creating Interactive Exhibits

Developing retro-featured themed events that emphasize the culture and history of the destination while creating retro-branding can help attract more tourists to the destination and increase tourism revenue. For this, festivals or fairs can be organized that depict the cultural and historical characteristics and advantages of the cultural heritage destination in a retro way. Creating interactive exhibitions or festivals with retro content can help visitors to experience both modern and nostalgic together.

Retro Design Elements

The use of retro design elements in marketing materials such as posters, brochures and merchandise to create a nostalgic and vintage feel in destinations is important in the context of holistic marketing of the destination in terms of retro marketing. The use of retro design elements is also important in the retro-branding process. Cultural heritage destinations that want to create a brand strategy with a long-term and retro content should act in a way that integrates traditional marketing practices such as social media posts, logo, billboard, slogan in their marketing communication. Thus, while creating a retro and original image suitable for the personality of the destination, it will contribute to the development of a brand personality that attracts visitors with this aspect.

Case Studies/Best Practices

Rebranding and retro-branding of cultural heritage destinations provides significant advantages to destinations. There are some destinations that enable successful use of the previously mentioned strategies and other strategies compatible with these strategies. For example, the industrial dominant image of the Spanish city of Bilbao has changed culturally with some practices carried out in the destination. The precision shown in the construction of the Guggenheim Museum in Bilbao and the holistic steps taken toward its rebranding have brought it to the forefront of the cultural characteristics of an industrial city, causing a large number of visitors to visit this city. In addition to redesigning the existing one to its original form, rebranding can also involve expanding the range of attractions and experiences that exist in the destination. For example, the Palace Museum in Beijing, China, has successfully used rebranding strategies to attract more visitors, improving the visitor experience and expanding the museum accordingly.

One of the good examples that can be given in rebranding is the transformation process of the city of Liverpool in the United Kingdom from an industrial city image to a cultural destination (Daramola-Martin, 2009). Through rebranding, Liverpool has increased visitor numbers and revenues by focusing and properly marketing musical and cultural heritage in a structure that appeals to the modern traveler. Against this rebranding example, Kowloon City, Walled Park (see Figure 9.1), which is affiliated with Hong Kong, can be given as an example of retro-branding. Accordingly, while Kowloon City exhibits the traditional architecture of Chinese culture and the historical importance of this region in Walled Park, it gives its visitors a sense of nostalgia and regains its former glory with the restorations carried out for this purpose (Lo, 2020).

Retro-branding and rebranding provide a balanced modernization, while emphasizing their cultural and historical significance for destinations. Additionally, retro-branding and rebranding can also involve the creation of new attractions inspired by the cultural heritage and historical richness of a destination. An example of this is the destination management of Havana, Cuba, reflecting the rich culture and historical structure of the city in line with the retro-branding, and designed with this integrity, and new bars and restaurants opened in parallel. Cultural Heritage management and rebranding/

FIGURE 9.1 Kawaloon City, Walled Park

Source: https://www.lcsd.gov.hk/en/parks/kwcp/index.html

retro-branding are used to increase the spending of visitors in England and to strengthen the brand image of cities. In Newcastle, for example, £1.6m was awarded from the Heritage Lottery Fund to rebuild and renovate the city's historic Bigg Market, which was used for retro-branding. At Southport, a contemporary lighting system is used for the trees, while at the same time restoring the tradition of the 1950s Southport tree lights to revive the forgotten heritage. Blackburn has produced classic bus tours and 'Talk of the Town Guides' Funded by the Heritage Lottery, while organizing history walks to help people discover and learn about their local heritage. In Bournemouth, however, it is aimed to contribute to the marketing of cultural heritage tourism and the branding of the country on cultural heritage by supporting events such as 'The Great Exhibition of Boscombe' in a region of many coastal market towns in order to develop different identities through heritage (Historic England, 2020).

Conclusion and Future Directions

Rebranding and retro-branding are two important strategies used to revitalize heritage destinations. Rebranding focuses on modernizing a destination to appeal to modern travelers, while retro-branding focuses on restoring the historical and cultural significance of a place. Both strategies aim to strike a balance between modernization and preservation of cultural and historical aspects of a place. However, it is essential to strike a balance between modernization and preservation of cultural and historical aspects of a place (Kumar et al., 2019). Overemphasis on modernization can lead to the loss of cultural and historical significance, while overemphasis on preservation can make the destination less attractive to modern travelers. Therefore, it is crucial to take a balanced approach that honors the location's cultural and historical relevance while still providing access to contemporary conveniences and experiences. Additionally, it's crucial to include local stakeholders in the rebranding or retro-branding process to guarantee that the adjustments are long-lasting and advantageous to the neighborhood.

Rebranding or retro-branding of historic locations is a successful strategy for drawing tourists and boosting sales. Using design styles from previous eras evokes a sense of familiarity between memories and destinations for tourists. However, it is essential to strike a balance between modernization and preservation of cultural and historical aspects of a place (Kumar et al., 2019). Heritage destinations that adopt a balanced approach can remain relevant and attract visitors while preserving their cultural and historical significance. The success of a rebranding or retro-branding strategy also depends on collaboration with stakeholders and local communities. Collaborating with stakeholders and local communities can help to ensure that the rebranding or retro-branding strategy is well-received, leading to increased support and revenue (Ratten, 2023). In addition to maintaining a delicate equilibrium between modernization and preservation, effective communication is essential for the success of retro-branding or rebranding initiatives in cultural heritage destinations. Clear and compelling communication helps to convey a site's unique story and connect the past and present for visitors. By presenting a narrative that highlights the destination's evolution of the destination while emphasizing its cultural and historical roots, heritage sites can establish a deeper connection with tourists. This communicative approach enhances the visitor experience and contributes to the overall success of the rebranding or retro-branding strategy.

The success of retro-branding or rebranding strategies for heritage destinations depends on a comprehensive approach. The balance between modernization and preservation, effective communication, sustainable practices and the integration of technology plays a crucial role in reviving these cultural gems. Through collaboration with local stakeholders and communities and using these approaches strategically, heritage destinations can not only attract a variety of visitors, but also ensure the longevity of their historical and cultural significance in the global tourism landscape (Palmer & Tivers, 2018).

Future trends should look to the future, rather than the past, wherever possible (Carson, 2018). Therefore, future trends should focus on creating a destination brand suitable for the sense of place and taking meaningful initiatives that reflect the unique identity of the destination. Heritage provides competitive advantage to destinations through rebranding and retro-branding by providing originality and reliability, expressing the culture of the destination, increasing the destination attractiveness and providing competitive advantage to the destination. In conclusion, the rebranding of heritage destinations and retro-branding strategies are becoming increasingly important in the tourism industry. Rebranding or retro-branding of heritage destinations is likely to play an important role in the tourism industry in the future. Branding strategies that reflect the nostalgic atmosphere of the past can attract tourists and emphasize the uniqueness of destinations. Leveraging a destination's past can help determine its future and increase tourist preference. Destinations using successful rebranding should leverage nostalgia to create a sense of familiarity and connection to their visitors. Destinations that can effectively capitalize on the nostalgia and familiarity of the past will likely establish a strong emotional bond with visitors and stand out from the competition. However, when implementing such a branding strategy, the protection of cultural heritage and sustainable tourism objectives should be considered. Also, digital marketing tools and social media can play an important role in the success of retro-branding strategies. By following these recommendations and staying up-to-date with current trends, destinations can successfully rebrand and attract new audiences while maintaining their heritage and identity. In the future, cultural heritage rebranding or retro-branding in the tourism industry may become more common as a marketing strategy that will attract tourists and emphasize the uniqueness of destinations.

References

Andriotis, K., Foroudi, P., & Marvi, R. (2021). Heritage destination love. *Qualitative Market Research: An International Journal*, *24*(2), 240–259.

Balmer, J. M. (2011). Corporate heritage identities, corporate heritage brands and the multiple heritage identities of the British Monarchy. *European Journal of Marketing*, *45*(9/10), 1380–1398.

Brown, S., Kozinets, R. V., & Sherry, J. F. Jr. (2003). Teaching old brands new tricks: Retro branding and the revival of brand meaning. *Journal of Marketing*, *67*(3), 19–33.

Carson, N. (2018). How to reawaken a brand's heritage. https://www.creativebloq.com/advice/how-to-reawaken-a-brands-heritage (Accessed 02.04.2023).

Chen, P. C. (2022). The effects of brand revitalisation and retro branding on brand and purchase outcomes: The moderating roles of consumer nostalgia proneness and self-construal. *Journal of Marketing Management*, *38*(7–8), 771–799.

Daramola-Martin, A. (2009). Liverpool One and the transformation of a city: Place branding, marketing and the catalytic effects of regeneration and culture on repositioning Liverpool. *Place Branding and Public Diplomacy*, *5*(4), 301–311.

Giudici, E., Melis, C., Dessì, S., & Francine Pollnow Galvao Ramos, B. (2013). Is intangible cultural heritage able to promote sustainability in tourism? *International Journal of Quality and Service Sciences*, *5*(1), 101–114.

Goi, C. L., & Goi, M. T. (2011, February). Review on models and reasons of rebranding. In International Conference on Social Science and Humanity (Vol. 5, No. 2, pp. 445–449).

Gülmez, M., Babür, S., & Yirik, Ş. (2012). Turizmde destinasyon markalaşması ve alanya örneği. *I. Disiplinlerarası Turizm Araştırmaları Kongresi*, *15*, 18.

Gursoy, D., & Mcleod, R. (2019). Marketing and branding heritage tourism. *Journal of Heritage Tourism*,*14*(8–9), 681–686.

Hargrove, C. M. (2002). Heritage tourism. *CRM-WASHINGTON*, *25*(1), 10–11.

Historic England (2020). Using Heritage in Place Branding. https://historicengland.org.uk/images-books/publications/using-heritage-in-place-branding/heritage-place-branding/ (Accessed 23.03.2023).

Huh, J., Uysal, M., & McCleary, K. (2006). Cultural/heritage destinations: Tourist satisfaction and market segmentation. *Journal of Hospitality & Leisure Marketing*, *14*(3), 81–99.

Kim, H., Stepchenkova, S., & Babalou, V. (2018). Branding destination co-creatively: A case study of tourists' involvement in the naming of a local attraction. *Tourism ManagementPerspectives*,*28*, 189–200.

Kockelman, K. M., & Lohr, S. L. (2019). Rebranding heritage tourism: A conceptual framework for creatively reimagining place. *Journal of Heritage Tourism*, *14*(8–9), 820–839.

Kumar, P., Siddiqui, A., Gupta, K., Jain, S., Bharath, B. D., & Maithani, S. (2019). Understanding urban environment in northwest Himalaya: Role of geospatial technology. In R. Navalgund, A. Kumar, & S. Nandy (Eds), *Remote sensing of northwest Himalayan ecosystems* (pp. 403–431). Springer.

Laws, E. (2007). Management of cultural and heritage destinations. In I. Yeoman, & S. Drummond (Eds), *Quality issues in heritage visitor attractions* (pp. 78–96). Routledge.

Li, X., Li, X. R., & Hudson, S. (2013). Destination branding: A review of the branding literature and proposed conceptual model. *Journal of Hospitality and Tourism Management*, *20*(1), 1–12.

Lin, L., Xu, Y., & Tao, Q. (2020). Motivational drivers of Chinese consumers' brand avoidance behaviours: A perspective of sportswear. *International Journal of Fashion Design, Technology and Education*, *13*(1), 45–57.

Lo, Z.. (2020). 10 unexpectedly revitalised historic sites in Hong Kong: Now and then. https://www.tatlerasia.com/homes/property/revitalised-historic-sites-hong-kong (Accessed 01.04.2023).

Murphy, J. (1988). Branding. *Marketing Intelligence & Planning*, *6*(4), 4–8.

Mutana, S., & Zinyemba, A. Z. (2013). Rebranding the Zimbabwe tourism product: A case for innovative packaging. *International Journal of Advanced Research in Management and Social Sciences*, *2*(4), 95–105.

Paddison, B., & Biggins, R. (2017). Advocating community integrated destination marketing planning in heritage destinations: The case of York. *Journal of Marketing Management*, *33*(9–10), 835–857.

Palmer, C., & Tivers, J. (Eds.). (2018). *Creating heritage for tourism*. Routledge.

Rahman, M. S., Abdel Fattah, F. A. M., Hussain, B., & Hossain, M. A. (2021). An integrative model of consumer-based heritage destination brand equity. *Tourism Review*, *76*(2), 358–373.

Ratten, V. (2023). Heritage entrepreneurship: Future trends. In V. Ratten (Ed.), *Heritage entrepreneurship. Palgrave studies in global entrepreneurship*. (pp. 153–166). Palgrave Macmillan.

Seraphin, H., Yallop, A. C., Capatîna, A., & Gowreesunkar, V. G. (2018). Heritage in tourism organisations' branding strategy: The case of a post-colonial, post-conflict and post-disaster destination. *International Journal of Culture, Tourism and Hospitality Research*, *12*(1), 89–105.

Su, D. N., Nguyen, N. A. N., Nguyen, Q. N. T., & Tran, T. P. (2020). The link between travel motivation and satisfaction towards a heritage destination: The role of visitor engagement, visitor experience and heritage destination image. *Tourism Management Perspectives*, *34*, 100634.

Wang, D., Li, X., & Liang, X. (2016). Rebranding a world heritage site: A case study of the Wulingyuan scenic and historic interest area, China. *Journal of Travel Research*, *55*(8), 1026–1039.

10

PLANNING HERITAGE TOURISM FROM A SUSTAINABLE MARKETING PERSPECTIVE

Alba Viana-Lora

Introduction

Tourism is one of the major industries worldwide, and its growth has generated a great impact on the economy, environment, society and culture of many communities (Deery et al., 2012; Gwenhure & Odhiambo, 2017). In this context, heritage resources play a key role in attracting tourists and preserving the cultural identity of tourism destinations (Nasser, 2003). Heritage resources are those tangible and intangible elements that have historical, cultural, artistic or scientific values, and are considered of interest to society, such as monuments, buildings, archaeological sites, museums, art collections, cultural events, popular festivals, traditions and customs, among others (Munjeri, 2004). All of them are a sample of a community's heritage and reflect its identity, values and history (Alberti & Giusti, 2012). Their importance in tourism lies in the fact that these elements are an attraction for visitors, who seek to learn about and experience the culture and history of the destinations they visit (Bonn et al., 2007). In addition, heritage resources are also a resource for the economic development of local communities, as their use for tourism can generate employment, encourage investment, increase commercial activity and improve the quality of life of the population (Jones & Munday, 2001).

Therefore, their preservation will be necessary for, on the one hand, the development of tourism in many regions and, on the other hand, the preservation of the cultural identity of the communities (Ezenagu, 2020). The protection of these elements makes it possible to keep traditions and cultural manifestations alive, to transmit them to new generations and to promote respect and appreciation of cultural diversity (Nocca, 2017).

DOI: 10.4324/9781003346197-10

However, their tourist exploitation can generate negative impacts, such as degradation, congestion and saturation of tourist sites, the disappearance of the authenticity and originality of cultural manifestations and the gentrification of historic districts (Hovinen, 1995; Zhuang et al., 2019). Responsible and sustainable management will be essential to avoid these negative impacts and ensure their long-term preservation (Millar, 1989). This implies the identification of heritage resources and their tourism valorisation, the definition of objectives and strategies for tourism development, the identification of tourists' needs and demands and the definition of management and control measures (Nasser, 2003). This makes it possible to satisfy current needs without compromising the satisfaction of future generations (Hunter, 1997).

One of the main challenges faced by heritage managers is the balance between conservation and generating economic benefits for local communities (Aas et al., 2005; Chong & Balasingam, 2019).

Some of the strategies pursued advocate promoting responsible and sustainable tourism. This implies the implementation of appropriate management measures, such as tourism planning (Millar, 1989), control of carrying capacity (Mota et al., 2021), diversification of the tourism offer (Datzira-Masip, 2006), education and awareness-raising of tourists (Ballantyne et al., 2009) and the involvement of local communities in the management of cultural heritage (Landorf, 2009).

Marketing can play an important role in this development, as it can influence tourist behaviour and the image they have of a given tourist destination (Ramkissoon et al., 2011). In particular, sustainable marketing contributes to the promotion of sustainable tourism practices and awareness of the importance of cultural and natural heritage conservation (Chhabra, 2009; Donohoe, 2012; Gilmore et al., 2007). More and more tourists are aware of the importance of sustainability in tourism and are looking for destinations that promote responsible practices in terms of environmental management (Buffa, 2015). Sustainable marketing can help meet this demand by promoting the use of renewable energy, waste reduction, consumption of local products, protection of heritage sites and promotion of sustainable tourism practices (Pan et al., 2018). It will also help generate income for local communities while respecting the environment (Cristobal-Fransi et al., 2020). Promoting local culture and offering authentic and respectful products and services will increase responsible tourist interest (Chan, 2010). This can help ensure the economic sustainability of heritage sites and promote the long-term conservation of cultural and natural heritage (Chhabra, 2009). For this reason, this study seeks to analyse the state of the art on sustainable heritage marketing from a supply side perspective, identify sustainable marketing management challenges and strategies and highlight research gaps in order to propose a future research agenda.

Method

This chapter presents a systematic literature review on sustainable marketing in heritage tourism places. The literature review allows for the compilation and analysis of available information on a specific topic, which provides insight into the most relevant developments and findings in the field (Carter et al., 2015). It also allows for the identification of research gaps, as well as areas where additional research is needed (Munn et al., 2018). Synthesising existing information can be useful for future research and decision-making in the field (Rother, 2007). It is a technique used before in the field of tourism and marketing (Han & Bai, 2022; Li et al., 2020; Saur-Amaral et al., 2012).

The Web of Science database was used for information extraction. This database is widely used by researchers for presenting high-quality research (Chavarro et al., 2018). The following search algorithm was used for the search:

touris* AND "heritage" AND sustainab* AND "marketing"

This allowed us to extract publications containing the words marketing, heritage and those derived from tourism and sustainability in the title, abstract, author keywords and Keywords Plus. We filtered by document type to select article, book chapters, review article, early access and book and by language to extract publications in English.

The inclusion criteria are set out in Figure 10.1. They allowed for the selection of English language publications of article, book chapters, review article, early access and book that dealt with sustainable marketing in heritage tourism from the supply side. Tourists visiting heritage sites are very diverse and there is no clear profile determined, so I consider that demand-side analysis should be addressed in other future research (Chhabra, 2009; Tarawneh & Wray, 2017). Articles that were not relevant to the topic of study and those that did not meet the inclusion criteria were excluded.

The systematic literature review consisted of three phases specified in Figure 10.1. The first phase was based on the search for information. The search result yielded 268 documents as of 20 February 2023. The second phase focused on a filtering process. The data was downloaded into an Excel file and the first filtering process was carried out by reading the title and abstract. This process resulted in 77 publications. A second filtering process was initiated in which the full text of each publication was read, eliminating 67 documents that did not have sustainable marketing in heritage tourism sites as their main objective. Therefore, ten documents passed to the third phase consisting of a content analysis. In this phase, an in-depth reading was carried out again to verify that all the articles met the inclusion criteria and

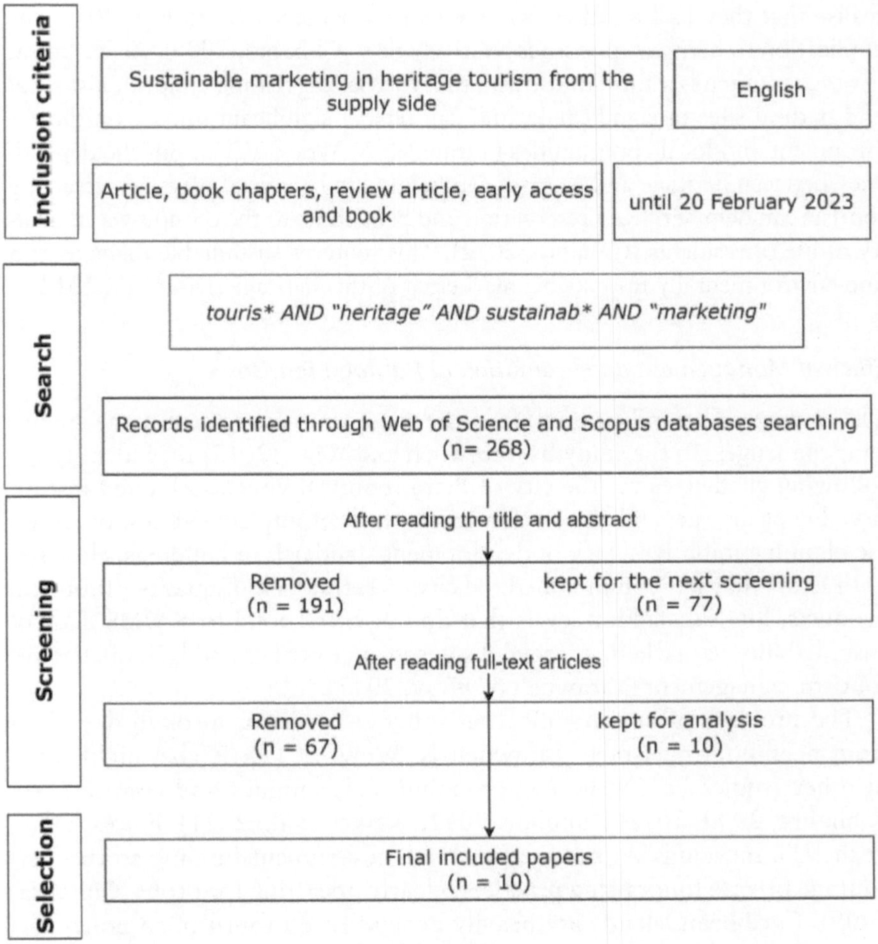

FIGURE 10.1 Inclusion criteria and stages of the systematic literature review process. Own elaboration

the content analysis was carried out, the results of which are shown in the following section.

Results

This literature review has highlighted the challenges facing the efficient management and promotion of heritage tourism and sets out sustainable marketing, stakeholder collaboration and brand building as key strategies to address these challenges.

Sustainable marketing emerged in the 1990s, when consumers became more concerned about the environment and sustainability, and businesses began to realise that they had to adjust to new market demands (Donohoe, 2012). Its application to heritage tourism is relatively new (Chhabra, 2009). As we know, heritage tourism is a form of tourism that focuses on visiting cultural, historical and natural sites of significance, and can have a significant impact on the environment and local communities (Tarawneh & Wray, 2017). But the dependency between heritage and tourism can be used in beneficial ways, as attracting tourists can help heritage preservation and contribute to the economy and quality of life of residents (Chhabra, 2012). This requires sustainable management and environmentally responsible marketing of this heritage (Donohoe, 2012).

Efficient Management and Promotion of Heritage Tourism

The management and promotion of heritage tourism present some significant challenges. In the study by Tarawneh and Wray (2017) they identify the following challenges for the city of Petra: political violence, limited institutional capacity, dependence on foreigners, lack of implementation of strategic planning initiatives, lack of development standards in buildings, close ties with Israel that has commodified the city of Petra, lack of quality of tourism products, low satisfaction levels that do not correspond to a WHS, lack of sustainability in marketing, social displacement of tribes and lack of effective tourism management (Tarawneh & Wray, 2017).

The problem of commodification suffered by Petra, through day visits from neighbouring Israel (Tarawneh & Wray, 2017), is also highlighted in other studies linking heritage tourism to economics and consumerism (Chhabra, 2009, 2012; Donohoe, 2012; Walker et al., 2021). For example, in the US, museums do not receive much government funding so they encourage private fundraising practices, clearly operating for profit (Chhabra, 2009). Caribbean islands are heavily dependent on tourism, so policy has generally been geared towards finding ways to increase visitor arrivals and has not placed sufficient emphasis on factors that promote sustainability (Walker et al., 2021). Designing marketing solely with economic objectives in mind will undermine heritage preservation (Chhabra, 2009). Caribbean islands saw economic benefits from tourism, but over time this led to economic imbalances (Walker et al., 2021). This commodification of heritage is linked to the overloading of heritage sites, as mass tourism can lead to saturation or overloading of space (Chhabra, 2009). This mass tourism breaks the rule of conservation of heritage sites (Donohoe, 2012).

These problems should be addressed from a sustainability perspective, with the aim of minimising the negative impacts of heritage tourism on the environment and local communities and ensuring the long-term preservation of cultural heritage (Donohoe, 2012). There is an immediate need for an integrative approach to the sustainable development, management and marketing of heritage sites (Tarawneh & Wray, 2017).

Sustainable marketing is the right tool for this management. However, sustainable marketing in tourism can be complex due to the large number of stakeholders involved in the industry (Donohoe, 2012). Strategic planning must be advocated to reconcile the often conflicting interests of various stakeholders and to ensure effective management and promotion of cultural heritage tourism.

Most marketing strategies are short-term (Chhabra, 2009) and sustainable marketing can help to preserve heritage in the long term by balancing profit and preservation (Chhabra, 2009). It will take into consideration social, natural and cultural aspects and current and potential impacts (Chhabra, 2012). Heritage tourism organisations need to consider where they are now, where they want to get to and how they are going to get there (Tarawneh & Wray, 2017). At the same time, aspects such as the overall mission, strategic planning and analysis of the environment that are rarely taken into account (Chhabra, 2009) need to be considered. To this end, Chhabra's (2009) sustainable marketing framework captures these issues and has value in the planning stages of the process (Donohoe, 2012). This sustainable management will require innovative products, a marketing plan that is sensitive to the ethical and moral and environmental demands of the consumer and the global community (Donohoe, 2012).

De-marketing or demarketing of heritage is a strategy within the sustainable marketing literature (Chhabra, 2012; Donohoe, 2012; Gregory-Smith et al., 2017; Tarawneh & Wray, 2017). It involves steering tourists away from areas at certain times or places (Donohoe, 2012). Such visitor management can preserve cultural heritage and ensure its sustainability in the future (Donohoe, 2012; Tarawneh & Wray, 2017). But decommercialisation to alleviate high tourist pressure is complicated because tourism activity generates economic benefits in the destination (Tarawneh & Wray, 2017).

One way to reduce tourism at heritage sites, as proposed with demarketing, while maintaining the economic benefits, would be to focus on slow tourism. Walker et al. (2021) support to slow tourism, focusing on the quality of the tourism experience rather than the quantity of places visited or activities undertaken, as an antidote to mass tourism. It seeks to promote more conscious and responsible tourism, allowing travellers to connect with culture, nature and local communities in a more authentic and meaningful way (Walker et al., 2021). Adie (2019) advocates community-based tourism, which seeks to maximise economic benefits for local communities and reduce the reliance of mass tourism on large tour operators.

Other strategies within sustainable marketing are found in the literature such as tourist education (Chhabra, 2009; Donohoe, 2012), the use of digital technologies that can help maximise economic benefits and minimise negative impacts on cultural heritage and local communities (Chhabra, 2022) or the diversification of tourism offerings (Tarawneh & Wray, 2017). But effective stakeholder collaboration and heritage branding deserve special attention.

Effective Collaboration between Stakeholders

Most articles highlight the need for effective collaboration between stakeholders such as governments, tourism businesses, local residents and indigenous communities, to achieve effective management and promotion of cultural heritage tourism (Banerjee & Mukherjee, 2022; Chhabra, 2009, 2012; Donohoe, 2012; Gregory-Smith et al., 2017; Tarawneh & Wray, 2017; Walker et al., 2021; Wells et al., 2015). The success of the sustainable marketing approach relies on collaboration between stakeholders (Chhabra, 2009). Collaboration between these stakeholders is important because it enables more effective and sustainable management of heritage tourism through the creation of sustainable tourism products, responsible heritage promotion, mitigation of negative impacts and increased benefit to local communities (Banerjee & Mukherjee, 2022; Chhabra, 2009, 2012; Donohoe, 2012; Gregory-Smith et al., 2017; Tarawneh & Wray, 2017; Walker et al., 2021; Wells et al., 2015). Donohoe's (2012) study confirms the importance of partnerships in marketing to manage the Rideau Canal in Canada. However, Chhabra's (2009) study highlighted that only one museum understood the importance of external environment analysis and only one museum sought to help local culture by partnering with trading posts, colleges and universities (Chhabra, 2009). This means that the museums analysed by Chhabra (2009) generally operate from individualism, without considering the factors or agents involved in tourism, although the author does highlight the need to create partnerships (Chhabra, 2009). Within collaboration, we could introduce the collaboration and involvement of staff working at heritage sites (Gregory-Smith et al., 2017; Wells et al., 2015), as employees could be considered a core target of behaviour change and instilling sustainability in employees influences the process of marketing sustainability messages on visits (Gregory-Smith et al., 2017; Wells et al., 2015). Sustainable practices need to be communicated to stakeholders and this could be done through training and environmental awareness programmes (Gregory-Smith et al., 2017).

Consolidation of the Heritage Brand

One of the aspects to consider within sustainable heritage tourism marketing is brand building. A region can transform its heritage brand to attract tourists and improve the local economy (Banerjee & Mukherjee, 2022). The most prominent example is the UNESCO World Heritage Site (WHS) designation. WHS is seen as a label that attracts international tourists (Chhabra, 2012), a unique brand (Tarawneh & Wray, 2017) and its heritage designation leads to a significant increase in visitors (Donohoe, 2012). The original intention was to safeguard the designated WHS sites, but recent motivations have shifted towards transforming them into a tourism product and a promotional tool

(Chhabra, 2012; Tarawneh & Wray, 2017). For example, with the declaration of the WHS, visits to the City of Petra increased from 93,000 in 1985 to 359,000 in 2006, a growth that continued when it was declared one of the seven Wonders of the World in 2007, reaching 975,000 visitors in 2010 (Tarawneh & Wray, 2017). However, this uncontrolled increase in tourists can have negative effects on the site, causing an environmental and cultural impact on the site, which can be detrimental to its long-term preservation (Chhabra, 2012; Tarawneh & Wray, 2017). Furthermore, World Heritage designation can also lead to gentrification of the area, which can lead to displacement and loss of cultural and community identity (Tarawneh & Wray, 2017). Therefore, brand building should be done through sustainable marketing.

Discussion

This study presents the need for sustainable management and environmentally responsible marketing of heritage tourism to minimise negative impacts on the environment and local communities, as well as to ensure the long-term preservation of cultural heritage. A number of challenges to the management and promotion of heritage tourism have been identified such as political violence, limited institutional capacity, lack of implementation of planning initiatives or commodification of heritage that need to be addressed through strategic planning and stakeholder collaboration. It is essential that heritage management organisations involve local communities in their operations. This can include working with local communities on conservation and sustainable development projects, as well as promoting local products and services. By working closely with local communities, organisations can gain greater community support and commitment, which can help ensure the long-term success of heritage management operations.

However, organisations responsible for heritage management need to have a clear understanding of visitor needs and expectations. This will enable them to deliver experiences that are engaging and meaningful, while maintaining sustainability standards. In addition, they should strive to reduce their carbon footprint and minimise their impact on the environment. This can be achieved through the adoption of sustainable practices, such as the use of renewable energy, proper waste management and the reduction of water and energy consumption. It is important that they promote the conservation and protection of biodiversity, as this can also attract visitors and improve the organisation's image. It should not be forgotten that this management should be transparent and accountable in its communication with visitors and other stakeholders. Transparency and accountability are also key to establishing visitor trust and loyalty, which can be an important factor in the long-term success of heritage management operations.

Future Research

Heritage tourism is a constantly growing and changing industry, and future research is essential to ensure sustainable and effective management. The review of the ten articles selected for this analysis provides an overview of current marketing strategies and management approaches in heritage tourism. However, there are still a number of areas that require further research. Key areas for future research in heritage tourism include the following:

1 Integration into local and regional development plans:
   - Investigate how heritage tourism can be effectively integrated into local and regional development plans.
   - Consider the needs and perspectives of local communities in decision-making processes.
   - Research how sustainable marketing can improve the quality of life for local communities while ensuring heritage conservation and sustainability.

2 Utilising new technologies and digital media:
   - Explore the effective use of emerging technologies, such as virtual and augmented reality, in heritage and sustainable tourism marketing strategies.
   - Examine how these technologies can offer immersive and educational experiences for visitors, increasing awareness and understanding of cultural heritage.
   - Investigate the potential benefits of geographic information systems in the planning and management of heritage resources.

3 Evaluating the effectiveness of sustainable marketing strategies:
   - Understand how heritage branding influences tourists' decision-making and perception of heritage tourism as a sustainable and responsible activity.
   - Assess the impact of marketing campaigns on promoting sustainability in heritage tourism.
   - Explore the adaptation of sustainable marketing strategies to different types of heritage and tourism destinations.

4 Collaboration between stakeholders:
   - Investigate how collaboration between various stakeholders can enhance the management of heritage tourism.
   - Explore effective ways for stakeholders to work together in addressing challenges and maximising the benefits of heritage tourism.

5 Role of education and awareness-raising:

- Study the role of education and awareness-raising in promoting sustainable heritage tourism.
- Examine how education initiatives can engage visitors and local communities in heritage conservation and promotion.

By conducting research in these areas, we can further advance the field of heritage tourism and contribute to its sustainable development and management.

Limitations

There are some limitations to this study that should be taken into account. The data used is based only on the Web of Science database and there may be relevant studies published in other databases or in specialised journals and books that were not included in the search. In addition, the selected articles may be biased towards certain approaches and perspectives. Most look at sustainable marketing from a Western and developed perspective, and alternative approaches and more diverse perspectives may have been overlooked. This may limit the applicability of the strategies in contexts other than the selected studies.

Conclusion

This chapter aimed to systematically review the literature on sustainable heritage marketing from a supply side perspective. This review has allowed us to identify the challenges and management strategies of sustainable marketing and to highlight research gaps in order to propose a future research agenda. In conclusion, the growing concern for the sustainable development of heritage tourism and its promotion through sustainable marketing strategies has been noted. The studies reviewed have highlighted the importance of careful planning and integrated management in the sustainable development of heritage tourism, and the need to involve all stakeholders in the process. Furthermore, the implementation of sustainable marketing strategies can contribute to increasing awareness and understanding of cultural heritage, attract more responsible visitors and encourage sustainable conservation of cultural heritage.

References

Aas, C., Ladkin, A., & Fletcher, J. (2005). Stakeholder collaboration and heritage management. *Annals of Tourism Research*, *32*(1), 28–48.

Adie, B. A. (2019). *World heritage and tourism: Marketing and management*. Routledge.

Alberti, F. G., & Giusti, J. D. (2012). Cultural heritage, tourism and regional competitiveness: The Motor Valley cluster. *City, Culture and Society*, *3*(4), 261–273.

Ballantyne, R., Packer, J., & Hughes, K. (2009). Tourists' support for conservation messages and sustainable management practices in wildlife tourism experiences. *Tourism Management*, *30*(5), 658–664.

Banerjee, S., & Mukherjee, P. M. (2022). Revamping heritage brand: A case of Murshidabad, West Bengal, India. *Place Branding and Public Diplomacy*, *18*(2), 190–200.

Bonn, M. A., Joseph-Mathews, S. M., Dai, M., Hayes, S., & Cave, J. (2007). Heritage/cultural attraction atmospherics: Creating the right environment for the heritage/cultural visitor. *Journal of Travel Research*, *45*(3), 345–354.

Buffa, F. (2015). Young tourists and sustainability. Profiles, attitudes, and implications for destination strategies. *Sustainability*, *7*(10), 14042–14062.

Carter, R. W., Thok, S., O'Rourke, V., & Pearce, T. (2015). Sustainable tourism and its use as a development strategy in Cambodia: A systematic literature review. *Journal of Sustainable Tourism*, *23*(5), 797–818.

Chan, J. K. (2010). Building sustainable tourism destination and developing responsible tourism: Conceptual framework, key issues and challenges. *Tourism Development Journal-An International Research Journal*, *8*(1), 24–32.

Chavarro, D., Ràfols, I., & Tang, P. (2018). To what extent is inclusion in the Web of Science an indicator of journal 'quality'? *Research Evaluation*, *27*(2), 106–118.

Chhabra, D. (2009). Proposing a sustainable marketing framework for heritage tourism. *Journal of Sustainable Tourism*, *17*(3), 303–320.

Chhabra, D. (2012). Management of world heritage sites: An integrated sustainable marketing approach. In de Pablos. P., Tennysson, R. & Zhao, J. (Eds.), *Global hospitality and tourism management technologies* (pp. 190–207). IGI Global.

Chhabra, D. (2022). Smart and sustainable marketing frameworks in heritage tourism. In *Resilience, authenticity, and digital heritage tourism*. Routledge.

Chong, K. Y., & Balasingam, A. S. (2019). Tourism sustainability: Economic benefits and strategies for preservation and conservation of heritage sites in Southeast Asia. *Tourism Review*, *74*(2), 268–279.

Cristobal-Fransi, E., Daries, N., Ferrer-Rosell, B., Marine-Roig, E., & Martin-Fuentes, E. (2020). Sustainable tourism marketing. *Sustainability*, *12*(5), 1865.

Datzira-Masip, J. (2006). Cultural heritage tourism—Opportunities for product development: The Barcelona Case. *Tourism Review*, *61*(1), 13–20.

Deery, M., Jago, L., & Fredline, L. (2012). Rethinking social impacts of tourism research: A new research agenda. *Tourism Management*, *33*(1), 64–73.

Donohoe, H. M. (2012). Sustainable heritage tourism marketing and Canada's Rideau Canal world heritage site. *Journal of Sustainable Tourism*, *20*(1), 121–142.

Ezenagu, N. (2020). Heritage resources as a driver for cultural tourism in Nigeria. *Cogent Arts & Humanities*, *7*(1), 1734331.

Gilmore, A., Carson, D., & Ascenção, M. (2007). Sustainable tourism marketing at a World Heritage site. *Journal of Strategic Marketing*, *15*(2–3), 253–264.

Gregory-Smith, D., Wells, V. K., Manika, D., & McElroy, D. J. (2017). An environmental social marketing intervention in cultural heritage tourism: A realist evaluation. *Journal of Sustainable Tourism*, *25*(7), 1042–1059.

Gwenhure, Y., & Odhiambo, N. M. (2017). Tourism and economic growth: A review of international literature. *Tourism: An International Interdisciplinary Journal*, *65*(1), 33–44.

Han, W., & Bai, B. (2022). Pricing research in hospitality and tourism and marketing literature: A systematic review and research agenda. *International Journal of Contemporary Hospitality Management*, *34*(5), 1717–1738.

Hovinen, G. R. (1995). Heritage issues in urban tourism: An assessment of new trends in Lancaster County. *Tourism Management*, *16*(5), 381–388.

Hunter, C. (1997). Sustainable tourism as an adaptive paradigm. *Annals of Tourism Research*, *24*(4), 850–867.

Jones, C., & Munday, M. (2001). Blaenavon and United Nations World Heritage Site status: Is conservation of industrial Heritage a road to local economic development? *Regional Studies*, *35*(6), 585–590.

Landorf, C. (2009). Managing for sustainable tourism: A review of six cultural World Heritage Sites. *Journal of Sustainable Tourism*, *17*(1), 53–70.

Li, Y., Dai, J., & Cui, L. (2020). The impact of digital technologies on economic and environmental performance in the context of industry 4.0: A moderated mediation model. *International Journal of Production Economics*, *229*, 107777.

Lin, H. C., Han, X., Lyu, T., Ho, W. H., Xu, Y., Hsieh, T. C., & Zhang, L. (2020). Task-technology fit analysis of social media use for marketing in the tourism and hospitality industry: A systematic literature review. *International Journal of Contemporary Hospitality Management*, *32*(8), 2677–2715.

Millar, S. (1989). Heritage management for heritage tourism. *Tourism Management*, *10*(1), 9–14.

Mota, L., Franco, M., & Santos, R. (2021). Island tourism carrying capacity in the UNESCO Site Laurisilva of Madeira. *Island Studies Journal*, *16*, 255–270.

Munjeri, D. (2004). Tangible and intangible heritage: From difference to convergence. *Museum International*, *56*(1–2), 12–20.

Munn, Z., Peters, M. D., Stern, C., Tufanaru, C., McArthur, A., & Aromataris, E. (2018). Systematic review or scoping review? Guidance for authors when choosing between a systematic or scoping review approach. *BMC Medical Research Methodology*, *18*, 1–7.

Nasser, N. (2003). Planning for urban heritage places: Reconciling conservation, tourism, and sustainable development. *Journal of Planning Literature*, *17*(4), 467–479.

Nocca, F. (2017). The role of cultural heritage in sustainable development: Multidimensional indicators as decision-making tool. *Sustainability*, *9*(10), 1882.

Pan, S. Y., Gao, M., Kim, H., Shah, K. J., Pei, S. L., & Chiang, P. C. (2018). Advances and challenges in sustainable tourism toward a green economy. *Science of the Total Environment*, *635*, 452–469.

Ramkissoon, H., Uysal, M., & Brown, K. (2011). Relationship between destination image and behavioral intentions of tourists to consume cultural attractions. *Journal of Hospitality Marketing & Management*, *20*(5), 575–595.

Rother, E. T. (2007). Systematic literature review X narrative review. *Acta paulista de enfermagem*, *20*, v–vi.

Saur-Amaral, I., Ferreira, P., & Conde, R. (2012). Linking past and future research in tourism management through the lens of marketing and consumption: A systematic literature review. In *Tourism management studies, printing, paper authored by Irina Saur-Amaral, Pedro Ferreira and Rosa Conde*, *9*(1): 35–40.

Tarawneh, M. B., & Wray, M. (2017). Incorporating Neolithic villages at Petra, Jordan: An integrated approach to sustainable tourism. *Journal of Heritage Tourism*, *12*(2), 155–171.

Walker, T. B., Lee, T. J., & Li, X. (2021). Sustainable development for small island tourism: Developing slow tourism in the Caribbean. *Journal of Travel & Tourism Marketing*, *38*(1), 1–15.

Wells, V. K., Manika, D., Gregory-Smith, D., Taheri, B., & McCowlen, C. (2015). Heritage tourism, CSR and the role of employee environmental behaviour. *Tourism Management*, *48*, 399–413.

Zhuang, X., Yao, Y., & Li, J. (2019). Sociocultural impacts of tourism on residents of world cultural heritage sites in China. *Sustainability*, *11*(3), 840.

11

THE CHALLENGE FOR TRANSFORMATIVE TOURISM OF DIFFICULT AND DISSONANT HERITAGE

Michael Fagence

> The ultimate objective [of tourism] is to generate/experience existentially authentic moments and ensure that these moments stimulate moral selving, resilience, and enhancement of the inner self by elevating the self to an awakening and transcendental state of mind.
>
> (Chhabra, 2021, p. 762)

Introduction

The backgrounding influence on this chapter is composed of three sources. One of these is a recent investigation that fits into the scope of battlefield tourism (Fagence, forthcoming). A second is a re-reading of Edensor's (1998) fascinating account of alternative interpretations of visits to the Taj Mahal. And the third is Chhabra's (2021) recent affirmation that an important, perhaps ultimate objective of tourism is for the experience of 'essentially authentic moments' to awaken the mind. These three influences contribute to the core of this study, of both its consideration of backgrounding influences and its inspection of them within the scope of a selected case study. It is important at this early stage to point out that the nature of this study, although conventional in that it is designed to expose the basic circumstances of a particular case by being both descriptive and explanatory, it is essentially exploratory in that it is called upon to address a situation about which there are competing viewpoints and interpretations. The case study is about the happenings of a particular episode in the history of Australia, an episode that is impacted by dissonant and similarly inclined interpretations and advocacies. These three influential sources underpin both the substance and the momentum of the study even though not one of them is dominant; other than that, Chhabra (2021) has provided a target against which the created knowledge base can be tested.

DOI: 10.4324/9781003346197-11

There is an expansive territory in the fields of research and practice of tourism that has as its focus matters of history and heritage, and, for the purposes of this study, there are linkages that draw in inspiration from expressions of telling stories that inhabit the broad domain of popular culture (see, for example, Lundberg & Ziakas, 2019; Palmer & Tivers, 2019). One nuanced area is based around telling stories through films, and some studies that have focused on that have provided useful directions for understanding how the telling of the stories have influenced the audiences and consumers of the often-dramatic presentations. Two such interpretations can be considered here. In her commentary about how film audiences react to and become associated with what they see on the cinematic screen, Landsberg has suggested that "at the interface between a person and a historical narrative about the past ... an experience occurs through which the person sutures himself or herself into a larger history" (2004, p. 2). In an earlier study, Sobchack commented that audiences "become involved in and understand the stakes in historical representation ... and see themselves not only as spectators of history but also as participants in and adjudicators of it" (1996, p. 7). What is distinctive in these two commentaries, and what can be drawn out of many of the studies of expectations of visits to battlefield sites (see, for example, Miles, 2016; Scates, 2016), is the tendency for the commentaries and interpretations to be prejudicial, emphasising the role of the selected 'good guys'. It was the contrast of this position with that taken by Edensor (1998) – with his three interpretations of reasons for visiting the Taj Mahal – that nudged the enquiry in this chapter towards negotiating a story-telling pathway that at least could negotiate what are difficult and dissonant differences of emphasis, perspective and interpretation. This is a core issue in considering how the experience of a confronting situation – such as a battle – may be transformative and contributory in some way to influencing a person's cognitive and emotional state of well-being (Chhabra, 2021).

The challenge for this study has been to follow through with the observations of Chhabra, Edensor, Landsberg, and Sobchack, and to probe the circumstances of differences of interpretation about an episode of Australian history in which a particular bushranger and outlaw (a type of highwayman) – Edward (Ned) Kelly – attracts both acclaim (as a folk hero) and opprobrium (as a criminal and villain), and to fit the circumstances of the episode to the transformative position outlined in the opening quotation from Chhabra. It is unlikely to be exceptional, but for the case considered in this chapter, three questions about transformativism may arise: is the experience of a tourist that engages with the story transformative if

- there was previously no particular persuasion about the story, but after encountering it, one of the two opposed positions (acclaim or opprobrium) is taken up – or, perhaps, no position is taken?

- as an outcome of the engagement, the position the tourist held before the encounter is changed (that is, if the tourist who formerly considered the outlaw to be a hero now considers him to be a villain, or *vice versa*)?
- as an outcome of the engagement, any previously held interpretation becomes confirmed, and may in fact become further entrenched?

Many commentators have wrestled with this problem of capturing 'an experience', and especially fitting it to descriptions of being memorable or transformative (see, for example, Duerden et al., 2018; Scott & Le, 2017), and in their particular enquiry, Kerstetter and Cho (2004) shone a light on the influence of preconceptions, prejudgements, prejudices, and prior knowledge that the consumer of the presentation may have held before engaging with a particular experience, and the willingness of consumers to change their previous mindset. In many commentaries, the objective is to move beyond the mere aggrandisement of a knowledge base and proceed towards the achievement of a new state of personal happiness and well-being (Kim et al., 2012; Sthapit & Coudounaris, 2018).

This chapter, firstly, engages in a brief encounter with four backgrounding issues: tourism and battlefield sites, experiences with difficult heritage environments, memorable tourism experiences, and transformative experiences. Secondly, it draws on a selected case study set in the context of battlefield tourism, with its complicated juxtaposition of diametrically opposed viewpoints about elements of the story being told. And, thirdly, it concludes with speculations about the conjuring of transformative experiences in situations of contrary interpretations, how they impact on the substance of presentation through the medium of heritage-based tourism, and what might be useful future directions of study.

Backgrounding Issues and Themes

Before reaching the case study in the next section of this chapter, there is a brief encounter with four backgrounding themes. For each of the themes, the subject matter is linked to experiences. The first of the themes sets the operational context of the story, battlefield tourism. This is followed by a brief exploration of experiences in difficult environments, especially where those difficulties are derived from alternative interpretations of the same resources. The third theme focuses on memorable tourism experiences, while the fourth draws the considerations more closely to transformative experiences and how they may contribute to personal levels of cognitive and emotional well-being.

Experiences and Battlefields Tourism

Although the study of battlefields and their contribution to tourism is a burgeoning field (see, for example, Lloyd, 1998; Ryan, 2007), there are three important challenges and limitations. One of these is that the most readily accessible literature in the English language is of studies of battlefields that were prominent places of conflict in the two major European wars of the twentieth century, with commentaries on the American Civil War of the nineteenth century positioned in second place. The scale and circumstances of these battles are not wholly replicated in other theatres of war. As for the second challenge, by far the majority of the published literature traces battlefield experiences of large-scale confrontations with large forces and casualty counts, extensive battleground sites, and confrontations that significantly changed economic, political, and social circumstances. And, thirdly, there is a degree of sameness in the way the post conflict circumstances have been analysed, with survey instruments being designed to elicit qualitative interpretations of what meanings the sites hold for various types of visitors (see, for example, Miles, 2016, and his earlier doctoral thesis; Scates, 2016, and his earlier doctoral thesis). Most studies of battles and battlefield sites exhibit a fascination for logistical details, for the provocations that led to the confrontations, and especially for any special military hardware and tactics.

At what may be termed a more human scale, some surveys focus on ranges of explanations for engaging in post-event battlefield experiences. Included among the responses of tourists to questions about their motivations for their visit are aspirations to engage with the 'sense' of the place, to imagine encounters with high profile combatants, to reflect about their own heritage and identity, to 'walk' where family members of earlier generations had walked and fought. For some visitors engagement with a battlefield site is not only an educative experience, but also a secular pilgrimage and an existential search for meaning (Dunkley et al., 2011; Hyde & Harman, 2011). In their assessment of secular pilgrimage to the Turkish enclave of Gallipoli (1915), Hyde and Harman have speculated about the personal constructions of 'sacredness', and 'deep spiritual meaning' of the site drawing on the imaginings of the various visitors to the site. A remembrance of the conflict is held annually at the Gallipoli site, and the bravery and commitment of both 'sides' of the conflict are marked in a service shared by representatives of the principal groups of national combatants. It is this commemoration of both 'sides' of the conflict that is a significant challenge.

Experiences and Difficult Heritage Environments

Perhaps sparked by the controversies that are an inevitable consequence of labelling an event or happening as 'dark tourism' there has been a substantial growth in interest and speculation about how to describe appropriately events, places and sites, and situations that have become targets for conjecture, dispute and dissonance, with consequential discrepancies in the interpretations of them. The scope of literature that is focused on heritage-based assessments, interpretations and studies is littered with terms such as contested (Silverman, 2011), difficult (Macdonald, 2009), dissonant (Tunbridge & Ashworth, 1996), uncomfortable, and with references to places of 'death and disaster' (Lennon & Foley, 2000), 'places of pain and shame' (Logan & Reeves, 2009), 'heritage that hurts' (Uzzell & Ballantyne, 1998), and so on. This diversity of descriptive interpretations should not be surprising because, as Tunbridge and Ashworth (1996) have claimed, heritage is controversial (p. 20), with differences of opinion and dissonance about heritage issues to be expected, and almost intrinsic to the nature of heritage discovery and interpretation. A recent review of published literature about contested heritage (Liu et al., 2021) reports that "given its wide range of stakeholders, different uses and need to balance preservation and change, it is unsurprising that heritage has become an arena of contestation" (p. 442), and it draws on Tunbridge and Ashworth's commentary to support the contention that disputes about difficult heritage environments are inevitable, especially as 'heritage', as a resource, is so susceptible to manipulation for cultural, economic, political or social reasons and purposes by various stakeholder groups. Disputations about which version of any story should be depicted for consumption through tourism, becomes an impediment to levels of cognitive and emotional well-being for any group and for any member of such groups.

Experiences and Memorable Tourism Experiences

Recent bibliographic and research reviews (see, for example, de Freitas Coelho et al., 2018; Duerden et al., 2018; Hosany et al., 2021; Hosseini et al., 2021; Scott & Le, 2017) share an interpretation that experience "is an elusive construct in the sense that there is remarkably little conceptual clarity about what exactly constitutes an experience" (Strijbosch, 2022, p. 2), and that this is a situation that is an outcome of involvement in the study of emotions, experiences and memories from a diversity of scientific disciplines, each with their niche perspectives. Each of the disciplines that becomes involved in these matters trade on their discipline-based intricacies, and each wrestles with a range of problems that are linked to timescale, the differentiation between short-term, real-time experiences, and the longer-term experiences that are memories. A further complication arises with speculations about 'prosthetic memories' (Landsberg, 2004). Another problem area, highlighted by Seyfi et al. (2019) in their study of cultural tourism and the experiences on offer, is the need for careful differentiation between those that are largely hedonistic, and those that are essentially eudaimonic, with the differentiation turning on the notion of what may be considered to be a meaningful and memorable experience, one that is "remembered and recalled and selectively reconstructed by the tourist" (p. 341). Certainly, the issues around tourism experiences are multi-faceted and difficult to untangle, leading to a situation in heritage-based tourism in which the same (or similar) set of circumstances are interpreted differently according to personally held viewpoints; this is the situation of the case study considered later. From recent research there have come suggestions about a movement towards a position described by de Freitas Coelho et al. (2018) as 'reflective engagement' in which a tourist's behaviour involves both cognitive and affective processing and learning, contributing to personal well-being (see also Pyke et al., 2016), and facilitating differences of interpretation of a consistent range of resources and happenings according to personally held preferences. While in the case study in this chapter those differences of interpretation are almost combative, in the Taj Mahal study of Edensor (1998), the differences between the three alternatives are largely compatible, a situation that is compatible with the commentary of Hosseini et al. (2023) that recollected experiences are usually unusual and extraordinary, and may entertain both positive and negative dimensions. Different interpretations may not be unusual; in the case study, they happen to be polarised.

Transformation and Transformative Experience

Why and how transformative experiences occur, and how and why individual tourists respond in particular ways, are questions that have exercised the cogitations and deliberations of many commentators. Many of the attempts to grapple with these issues have been captured in the bibliographic reviews of various aspects and expressions of the nature of experiences in, of and through tourism. Among the more recent reviews are those by, for example, Hosany et al. (2021), Hosseini et al. (2021), Nandasena et al. (2022), Teo et al. (2021), and these have been buttressed by interpretive commentaries from Chhabra (2021), Duerden et al. (2018), Pung and Chiappa (2020), Sheldon (2020), and others. Additional and complementing expositions include those that focus generally on tourism experiences (such as Scott & Le, 2017; Strijbosch, 2022), others that give particular attention to memorable tourism experiences (such as Hosany et al., 2021; Sthapit & Coudounaris, 2017), some that concentrate attention on transformative experiences (such as Chirico et al., 2022; Hoggan & Finnegan, 2023; Rus et al., 2022), and others that concentrate on religious tourism (Collins-Kreiner, 2010), on contested heritage (Liu et al., 2021), on quality-of-life indicators and wellness tourism (Sirgy & Uysal, 2016; Voigt et al., 2011).

Even this abbreviated list of selected sources can only help to gain a toehold on the nature and use of theory about experiences in, of and through tourism, and the potential implications of transformism. One of the reasons for this is the entry into considerations about tourism from strands of thought with their associated tools from a diversity of disciplines that have as their base camps philosophical, theoretical, and operational scopes that are considerably different from those of tourism. This creates possibilities of missing the cognitive and technical challenges peculiar to tourism, with adoptions missing out on necessary adaptations. Transformation and transformative experiences is one of the areas that is a particular challenge for tourism. The original purpose of transformation theory was to cope with formal educational processes and especially adult learning (see, for example, Mezirow, 1991; see also Chirico et al., 2022; Hoggan & Finnegan, 2023), with a learning process that was formal and with particular learning and assessable objectives; 'learning' and contributing to well-being in tourism is less structured and personally configured. It is, it can be suggested, more difficult to assess behavioural, cognitive, and emotional changes because of experiences in tourism; that is not to say that it cannot be done, nor that it should not be attempted, but that tourism presents a different challenge.

Issues of this kind are commented upon in the concluding remarks in the research agenda listed by Nandasena et al. (2022), and by Taylor and Cranton (2013). But, as Kirillova et al. (2016) have argued, consumers as tourists engage in actions and behaviour that reach out to personal growth through active participation, and through the achievement of improved levels of cognitive and emotional meaning they contribute, eventually, to new benchmarks of beliefs, ideologies, insights, and values (see also, Taylor & Cranton, 2013). The key, for tourism, is not the challenge of 'testing' through a formal educational process of examination, but through the willingness of tourists to draw on their powers of critical thinking to 'transform', to change (or even to consolidate) their previously held position (Rus et al., 2022; Voigt et al., 2011). Sheldon's (2020) position is that the usefulness for tourism of the educational theory of transformation lies principally with its capacity to open an avenue of awareness of what may contribute to the development of an individual's emotional, intellectual, and spiritual being, by setting the individual tourist in a context that is greater than themselves (see also Soulard et al., 2019). At this stage, even the substantial repertoire of bibliographic reviews does not confirm whether transformation is dependent upon a peak experience, or a dramatic revelation, or whether it can be the outcome of a lapse of intransigence or, conversely, a confirmation of previously held views (Zhao & Agyelwaah, 2023).

It is to gain some insight from the opportunities and challenges considered in this section that this narrative turns now to its case study.

Case Study

Methodology

The foundational elements assembled for this study are based on a programme that was established almost two decades ago to support doctoral research that grappled with the telling of stories about folk heroes from history through the medium of heritage-based tourism (Fagence, 2011). From that particularised information baseline, elements of the study have been regularly updated through conventional desk-based research and by 'muddy boots' fieldwork excursions in order to support a range of investigations that have used the core study as the central point from which to test a number of speculations about the telling of stories from history through tourism. This study has a particular niche in that it is positioned at the confluence of contentious considerations about historic battles, the positionality of even acknowledging them, and the challenges occasioned by the diversity of personal and community-wide reflections about them. For the case considered in this chapter, the positionality of the author/researcher has been steadfastly fixed on the authenticity and accuracy of what was being considered, even though foraging through the intricacies and nuances of positive psychology (see, for example, Filep & Pearce, 2014; Nawjin & Filep, 2016; Seligman, 2002; Sirgy & Uysal, 2016) became an inevitable task in order to grasp the linkages between motivations to visit and the securing of personal pathways to enrichment and enlightenment (see Fagence, 2023). For this research, it was never intended to engage in an extensive survey of visitors to the site of the battle; instead, there was resort to an invented form of autoethnographic investigation that could be undertaken by the author at the Stringybark battle site involving aspects of both actual and virtual participant observation, photography, site journal recording, and both on-site and off-site self-reflection (imagining about 'being there' without actually 'being there' at the time of the event). This 'invention' drew from the experiences of similar inventions used by Robinson (2015) and by Wall (2016), while reflections on positionality discoverable in study areas such as archaeology and even art history have been 'imported' into this study (Fryer, 2020) to link with reflections about the practice of ethnography (Palmer, 2009) so as to provide a substantive theoretical and conceptual basis for the approach taken in this study.

The investigative process used here has drawn on the advocacies about using a case study-based approach for bringing together elements that were not only descriptive and explanatory, but also exploratory (Yin, 2003, 2018). For the present study, the need to fuse together elements that were sustained by conflicting viewpoints was crucial, especially as the attractiveness for tourism of both the contextual story and the immediate circumstances of the historical event is dependent on either a resolution of the contesting viewpoints, or an acceptance that both can be sustained. The advocacies of Yin (2003, 2018) have been important in working through the process, as has the general advocacy of Moustakas (1990) with his observation that "each research process unfolds in its own way ... The heuristic researcher constructs methods that will explicate meanings and patterns of experience relevant to the question" (pp. 43, 44), and the commentary about loosening the rigours of conventional approaches to case study investigations recorded in de Urioste-Stone et al. (2018).

Following a synopsis of the story of Ned Kelly, and the unfolding of the battle at Stringybark Creek, the commentary here links the conjuring of potentially transformative experiences in 'parcels' that conform with the structure and sequence of the backgrounding issues considered previously. The case being put is not definitive, that is, it does not answer specific research questions, nor does it offer specific outcomes. Rather, it follows the encouragement of Mitchell (2006) and Yin (2003, see pp. 39–44, and later) to use the case study form as a means of explanation and experimentation, without necessarily seeking opportunities through other studies to achieve corroboration. This is addressed in the Conclusion.

Backgrounding

To help with an understanding of the peculiarities of this case, a brief outline of the Ned Kelly story is offered here. Although the particular episode that engages attention here was brief (a matter of a few hours) it represents an important trigger point that set in train a disjointed process that led eventually to what has been described as the 'last stand of the Kelly Gang', the siege at Glenrowan. The gunfight at Stringybark Creek is a small 'peak' event in what has been referred to as 'the Kelly Outbreak' (McQuilton, 1979), and it represents one small step in the nineteenth century of progress of Australia being an outpost of the British Empire to achieving the status of independence (1st January 1901). Although small in scale, the episode at Stringybark Creek is an outcome of the interplay of various economic, political, and social factors that were manifest in rural and remote districts especially of a distrust of the entrenched agencies of government that cumulatively led to a number of expressions of grievance and rebellion; the Kelly Outbreak is one of those expressions, and the Stringybark Creek 'battle' is one of the key triggers to an evolving profile of Australian history in a slow-burning period of angst towards and distrust of the agencies of government.

It was the confrontation at Stringybark Creek that catapulted the members of the Kelly Gang, and the leader Ned Kelly, into history as token rebels; the story of the confrontation has always been contested. One interpretation of the events confers hero status on Ned Kelly, based essentially on the stand he took against the widely despised agencies of government; but there is a contrary interpretation that he should be considered an outlaw and a murderer of policemen engaged in carrying out their lawfully backed duties. And it is this division of opinion that persists, and is the backgrounding for the dissonance about the actions at Stringybark Creek. There is little dispute about what happened at the site; the differences lie almost solely in the justification for any of the actions of the combatants. And it is this division that prompts questions about what transformations might be the outcome of encounters with the site in the 'now' taking into account what happened in the 'then'.

In the brief sections that follow, the episode of Stringybark Creek will be 'tested' against the four major themes considered previously.

The Battlefield Experience

The most usual case is for battlefield studies to focus on large-scale confrontations, with massed opposing forces in the hundreds or even thousands of combatants, with eventuating casualties also in large numbers, and engagements extending across several episodes and large expanses of territory; but, not all accounts are of events of such magnitude, and the Stringybark Creek episode fits most easily into the category of small civil disturbances such as scuffles and gunfights involving only a handful of adversaries. Two of the important issues in every case is what was the cause of the confrontation, and whether that cause is an isolated aggravation to a few people or a more extensive problem that impacts a much larger population; the Stringybark episode has been highlighted in the account of Australia's history more because of the involvement of Ned Kelly than because of the cause of which it was a symbol.

As a confrontation it was small-scale, with opposing forces of a police patrol of four men and the Kelly Gang also of four men. It was a gunfight, conducted through two phases, one at the police camp site, and the second 'somewhere' in the unkempt bush setting of small streams, massive stands of trees, a thick and almost impenetrable undergrowth, an irregularly undulating forested floor, no permanent buildings, and personally created tracks rather than any formed routeways. There are few markers that remain of any of the happenings on the site; the best that even the government-sponsored information panels can manage is a description that informs that 'this site is that which is most likely'. As the levels of accuracy and authenticity are so fragile, it has become a site where imagination, speculation, and intelligent guesswork contribute to an atmosphere of mystery and intrigue. It is a challenging site on which questions are raised about the locations of the police and Kelly Gang camp sites, how the drama unfolded, where the protagonists were positioned at any time during the gunfight, in which directions the protagonists moved during the confrontations, who was engaged in which actions at any time during the gunfight, the precise positions where the three policemen were attacked, shot and killed, the direction in which the only surviving policeman fled, the sightlines of the various episodes in the short gunfight, and what the Gang members did after the gunfight and in which direction they left the site. None of this action was recorded accurately at the time. None of it was scripted and pre-planned. The record of the surviving policeman and the final letters of Ned Kelly are the eyewitness accounts, and, as might be expected, the recollections of these foes have little that is in full accord.

"All of this confusion leaves the potential tourist with latitude for a thought-provoking, challenging, but also an enriching and enlightening encounter with the site" (Fagence, 2023, p. 163). The site is oppressive and threatening, enclosed by massive stands of eucalyptus trees, with sightlines across the site seriously impeded by thick undergrowth. In recent times, a rudimentary pathway system that supposedly links some of the sites of events during the gunfights has been constructed, and it is accompanied by information panels at or near positions where particular actions are thought to have taken place. So, for the purposes of tourism, what are the experiential options? Three themes can be identified. One of these could have a pilgrimage theme, in which the Stringybark Creek site can be included with others as a contribution to telling the Ned Kelly story, recognising his iconic and legendary status. A second, and perhaps companion theme, could be to link the action at Stringybark Creek with the final 'battle' at Glenrowan, associating both with the nineteenth-century movement that led eventually to the creation of the new nation of Australia, with the 'battles' contributing to the forging of a distinctive Australian identity. The third theme could be directed to recognising the valour of the police patrol, three members of which were killed during the gunfight exchanges, with the site being a focal point of memorialising the action of the police rather than the Kelly gang.

This site has strong emotive rationales for commemoration; the essential experiential problem is what is the key focal point – recognising the actions of the outlaw gang, or of the police patrol, or more broadly, the march of Australia towards independence? Each of these options has contestable elements.

Contested Rationales

In a previous section, it was suggested that contention of any kind about episodes of history and their heritage linkages arises because of the different versions of the story that are told, and which of those different versions is selected for depiction through tourism. There is an inevitability that situations told through stories about events in history will be targeted by conjecture, by dispute, and by dissonance of various kinds as the accounts differ, and the interpretations of them differ. This is the case with the Ned Kelly story in general, and the Stringybark Creek episode in particular. The particularity of the Ned Kelly story is identifiable with Tunbridge and Ashworth's (1996) claim that heritage is controversial; it is the subject of dissent about justifications for actions taken by both the Kelly Gang and members of the police patrol, even about questions such as 'who did what to whom and why, and how?', with agreement being difficult to achieve even about the precise locations of particular actions during the confrontation. Memorialisation of the action is challenging because there are two clearly differentiated and entrenched interpretations, and despite recent rapprochements between descendants of the principal families, the two interpretations prevail, often fuelled, and fanned by commentators in 'new' documentary discoveries. The various descriptions referred to previously about contested heritage, difficult heritage, and uncomfortable heritage, have a reasonable resonance for the site, not least because the Stringybark Creek episode has both appeal and meaning for different stakeholder groups, and the memorialisation of the action presents a challenge for each group.

This is a case for which the only solution seems to be for those who encounter the site to take cognisance of the surroundings, and try to fit the action into those surroundings, while also trying to empathise with any one of the principal viewpoints. This is not a case that is too similar to Edensor's (1998) conceptualisation of three viewpoints at the Taj Mahal; there, the three alternatives were largely compatible, whereas at Stringybark Creek, that is not so.

Memorable Tourism Experiences

This case study is a little different from many tourism experiences on several counts. Firstly, its scale is small; secondly, although it is a triggering event and on its own it may constitute a peak in interest in the Ned Kelly story, it is not, in fact, a peak of great magnitude; rather, it is a stumbling block at which two different interpretations meet. Any memorable experience taken away from a visit to the site will vary according to which interpretation of the story is held, and that might have been conditioned by preconceptual expectations, predispositions, and impressions. There is no extant available research that helps to establish what tourists 'do' with what they experience at the site, whether it confirms previously held views or changes them, whether it contributes to a better-informed position, whether it moves the visitor in any direction or to any degree along the pathway to personal improvement and values they hold. Similarly, there is no available information on whether visitors take away with them 'prosthetic memories', interpretations of being at the site when the happenings occurred (Landsberg, 2004).

With the absence of any published information, and without the resources to undertake a full scientific study of opinions of visitors to the site (to compare with many of the battlefield studies), the author set out to re-visit and experience the site for himself, adopting as far as possible an unprejudiced position set between the emotive positions supportive of either the police patrol or the Kelly Gang at the time of the happening. This was essentially an exercise of 'reflective engagement' (de Freitas Coelho et al., 2018), and it is a simplified and attenuated version of that used by Edensor (1998) with his study of the Taj Mahal. The intention was basically similar to Edensor, to identify what tourists could do at the site, to estimate how (and what) interactions there could be from different points of view, and to conjure at least some idea of what exposure to the site might do for their experience (leading, perhaps, to transformation – see later). The personal experience (of this author) conjured a number of outcomes: firstly, that there was the satisfaction of having found the site (it is remote), of having traversed it, and of gaining some respect for what the protagonists at the 'battle' would have encountered. Secondly, the encounter with the site generated a range of 'deeper meaning' experiences, not only in awe of what the circumstances might have been like in the late nineteenth century, but about bravery and cunning, about prowess and skill, about engaging in actions that (for the combatants at the time) had purpose and consequences, about commitment, about mateship (the legendary marque of Australian social behaviour) and sacrifice. It is this second level of experiences that situates the encounter with the site of Stringybark Creek as a meaningful experience, potentially persuasive of forming entrenchable interpretations about the historical combative encounter at the site.

Transformation and Transformative Experience

Of the four backgrounding issues considered in this section, this is the one that raises the biggest challenge. The bibliographic reviews listed in previous paragraphs and the studies of battlefield sites that were consulted (see especially Miles, 2016; Scates, 2016; and the doctoral theses of both) have not been particularly helpful for determining what, if any, transformation took place as a result of encounters with marquee and symbolic sites. That is not surprising as most studies never claim to be looking for that outcome.

As the site in this study is remote, as it is unlikely to be a target for serendipitous visitation, as it is unlikely to be visited by anyone that has not already formed some opinion about the circumstances of the confrontation, and, as far as has been established so far, it is a site that has not been subjected to a visitor survey, it remains conjectural whether visits have been transformational. Anecdotal comments posted to Ned Kelly web sites refer to the mysteriousness of visits to the site, and particularly to the problems that arise from uncertainties about precisely where action took place. Because of this, what follows in this section is a personal interpretation; it is incapable of being substantiated or refuted without a suitable and carefully constructed and administered study.

The situation of the site, its configuration and its landscape characteristics were described briefly in a previous section. What is gained from exposure to it is the almost impenetrability of the forest and the ground cover. This remains so despite occasional forest fires and some tree clearing exercises, and it poses questions about sight lines for gunfights, for the clearance of sites for camping, for the sustainability of pathways and tracks, for the accuracy of positioning sites for the gunfights and so on. Two sets of four men had to negotiate an uncoordinated disposition of stands of trees, a thick undergrowth, a network of streams and gullies, a number of disguised prospecting pits and tunnels, and a patchwork of cleared zones within which to find each other and their adversaries, to hide and prepare for 'battle', to move and circulate, and to engage in short spells of gunfights. There are no precise records of where the two principal gunfights took place, other than references to the first being on the perimeter of the police campsite, and the second 'somewhere' on a track close to the campsite. This is not helpful for developing a consciousness of where the really dangerous activity took place. However, if the information plaques can be believed, whereas the first was in a clearing in the forest, the second was almost anywhere in the bushland close to the camp site. Because of the oppressiveness of the forest, even now, the sites are imbued with a sense of mystery and sinisterness. As for the exit route taken by the one surviving policeman as he escaped, there are clues as to where that might have been, but that is as uncertain as the sites where the two patrolling officers were ambushed, and, after an exchange of gunfire, the first shot and killed on the spot and the other chased further into the dense bushland before being caught and shot. In addition, there is little certainty about what the four members of the Kelly Gang did after the gunfight, the route they took out of the Stringybark Creek site, and even the time of that exiting, and the intended destination.

So, what can be learnt for an encounter with this site? There is more meaning from participation through a visit to the story elements than a mere reading about them. A site visit contributes to new benchmarks of awareness and appreciation, of knowing or adding some specificity to speculations, of gaining a sense of the tension in the forest through two separate instances of gunfights of developing some empathy with the combatants, of contributing some purpose to the happenings. Sheldon (2020) has referred to 'a unique awakening experience'. But whether the experience of exposure to the site of action, and any personal conceptualisations or imaginings about the action will bring about a transformation is difficult to judge. Firstly, there is little hard evidence for this site and this happening to which reference can be made; any judgements committed here are conjectural, and somewhat conditioned by prior expectations based upon previous study of published sources. Secondly, what can be judged from the telling of the story through published mediums and films is that there is an ever-changing point of balance between the position of the Kelly Gang as outlaws that encountered deprivation and intimidation to become folk heroes, and sympathy for a small police patrol that was surprised, attacked, and fought back as best it could at the cost of three lives.

There is not a single target around which transformation can form; a visit to the site can help with providing experience of the site, and on that can be superimposed what is known about the story of the gunfight. Engagement with the site, and the support of published commentaries (before and/or after the visit) may generate a position of greater knowingness and empathy with the causes fought. Transformation of any kind (that is for or against the Kelly Gang) will be influenced largely by personal pre-conditioning, and by personal propensities to make a change of any sort. Suturing oneself (using Landsberg's proposition) to the site and the likely happenings is an inevitable challenge, and imagining the happenings further binds the visitor to the site to one or other of the viewpoints (pro or con the Kelly Gang/police patrol). But it does not help with differentiating between the opposing viewpoints, nor does it necessarily confirm the appropriateness of one over another. Even after several visits over a period of more than twenty years this author has not been persuaded to lean more heavily towards one viewpoint rather than another. Perhaps Sheldon's (2020) suggestion that the benefit of transformation theory is that it offers its capacity to open an avenue of awareness that might then facilitate the development of an individual's emotional, intellectual, and spiritual being by encouraging that individual to see the bigger picture, inclusive of many interests, hits the target.

Conclusions, Leading to More Questions

The case put in this chapter has been that for any situation in which a story from history is being told through tourism there may be occasions in which more than one interpretation of the circumstances in that story are feasible and supportable. An interesting revelation of this occurs in Edensor's (1998) fascinating commentary on alternative interpretations of meanings that can be associated with the Taj Mahal. He referred to them, not as different stories, but as different 'narratives', with each revealing different, contrasting and even competing interpretations, and with each embodying a significant symbolism. In his study he identified a range of narratives, each with their own associations. His interpretation of this situation was that many monuments and sites of historical interest would hold particularised interest that could be transferred into tourism attractions, with many embodying multiple narratives. Edensor's case study has a number of useful conceptualisations, interpretations, and models with resonance to what is considered here. The essential problem, however, is to gauge the transformative power of visits to historically important sites and features, and whether the transformation is triggered by the initial encounter, or whether it takes the form of a temporary state of well-being in advance of further encounters with the site or by indulging in further investigations. For the case presented in this chapter, the challenge is to work through the possibility that the transformation is manifest through a deepened commitment to the initial impression (and the state of well-being that is realised by it), rather than a change of support to a parallel narrative; in the case study, this may be sustained support (or even a deepened commitment) for 'Kelly as hero' rather than 'Kelly as villain' rather than a transfer of allegiance from one to the other – or, of course, *vice versa*. In each case, either the deepened experience or the change of support will rely on what was described by Chhabra as "elevating the self to an awakening and transcendental state of mind" (2021, p. 762). Neither Landsberg (2004) nor Sobchack (1996) offers any concrete evidence of transformation, and even Edensor (1998) positions his interpretation within a framework of multiple alternatives. An important question arises: 'to whom does the flow of transformation matter'? Is it an issue for visitors to a controversial site? or is it of consequence to stakeholders of the story, the site, and of any material evidence? Can there be an even-handed strategy of presentation that will tell the story without disturbing the balance of feeling about it? How might multiple and differentiated stakeholders manage the telling of the story so that its problems and dissonances do not hinder its presentation. Such questions are faced in the presentation of battlefield sites as tourism 'attractions' (where what happened there would most likely have more than a single transformative influence). Edelheim (2015), in a less specific context, provides some help here by considering destinations and activities as 'attractions' and interpreting them as mechanisms for transformation,

In order to understand the intricacies of the transformative process, this study considered four selected backgrounding issues. The first of these – battlefield tourism – was necessary to set the context in which the study was being conducted; the second focused on some of the intricacies of experience, and this was carried forward into a third issue, with its own focus on memorable tourism experiences; the fourth was an exploration of the nature of transformative theory and how that had a relevance for tourism. Then, these four issues were associated with the selected case study in order to expose the influence they could have on the presentation of a particular story from the history of Australia through tourism. What this study has found is that there is a vast spectrum of published literature that addresses the many perspectives that may be applied to the study of even the special case selected for consideration here, and that that spectrum is constantly being revised and augmented as special insights can be added from 'new' strands of thinking imported from an increasing range of disciplines. While much of the literature is repetitive, and generally supportive, some, and especially from the 'new' sources, is challenging. What persists, however, is a need for clarity about such matters as, for example, whether tourists seek transformation or even prepare themselves for it, whether there are degrees of transformation, whether transformation is dependent upon there being a peak experience, whether the sutured experience nominated by Landsberg (2004) is as relevant to tourism as it seems to be for cinema, whether Sobchack's (1996) references to prosthetic memories are as susceptible to transformation, whether multiple symbolisms embedded in stories from history are an impediment for the potential for a comfortable allegiance to any of them. A nagging series of questions remain, and they include, for example, what is the evidence of a transformative experience? can such an experience be repeated, and if so, does it lead to further entrenchment of a particular experience and interpretation? Remembering that it is not only the tourist who can be transformed, what might be needed are marketing-based strategies to encourage transformative experiences for tourists, with these becoming the catalyst for the development or even regeneration of a tourism attraction or destination. For the circumstances of situations such as the case study – with its multiple narratives – the challenge is as acute for the tourist as it is for the host site or destination (Sheldon, 2020).

So, what is the contribution of this chapter? Firstly, and drawing down from both Landsberg (2004) and Sobchack (1996), it considers a case in which the degree to which tourists visiting heritage sites become 'transformed' is a persistent challenge. Data about confirmation of positions or transformation of positions is scarce; where is there evidence of either confirmation or transformation, and what form does it take? Secondly, presumably, transformation is multi-directional in that it may substantiate a position or it may contradict it, and whichever it is, the outcome may be rational or it may be persuaded, or induced, or solicited through clever presentation. Thirdly, does it really matter, and if so, to whom? It is unlikely that an even-handed presentation of a challenging story would be attractive to all potential visitors, to planners, developers, and marketers; it would lack challenge and controversy. But … who knows? who has asked? This remains an under-researched and heavily nuanced field of study.

References

Chhabra, D. (2021). Transformative perspectives of tourism: Dialogical perceptiveness. *Journal of Travel and Tourism Marketing*, *38*(8), 759–768.

Chirico, A., Pizzolante, M., Kitson, A., Gianotti, E., Riecke, B., & Gaggioli, A. (2022). Defining transformative experiences: A conceptual analysis. *Frontiers of Psychology*, *2022*(13), Article 790300.

Collins-Kreiner, N. (2010). Researching pilgrimage: Continuity and transformations. *Current Issues in Tourism*, *19*(2), 1185–1189.

De Freitas Coelho, M., de Sevilha Gosling, M., & de Almeida, A. (2018). Tourism experiences: core processes of memorable trips. *Journal of Hospitality and Tourism Management*, *37*, 11–22.

De Urioste-Stone, S., McLaughlin, W., Daigle, J., & Fefer, J. (2018). Applying case study methodology to tourism research. In R. Nunkoo (Ed.), *Handbook of research methods for tourism and hospitality management* (pp. 407–427). Edward Elgar.

Duerden, M., Lundberg, N., Ward, P., Taniguchi, G., Hill, S., Widmer, M., & Zabriskie, R. (2018). From ordinary to extraordinary: A framework of experience types. *Journal of Leisure Research*, *49*(3–5), 196–216.

Dunkley, R., Morgan, N., & Westwood, S. (2011). Visiting the trenches: Exploring meanings and motivations in battlefield tourism. *Tourism Management*, *32*(4), 860–868.
Edelheim, J. (2015). *Tourist attractions: From objects to narratives*. Channel View Publications.
Edensor, T. (1998). *Tourists at the Taj: Performance and meaning at a symbolic site*. Routledge.
Fagence, M. (2011). 'Dead Men Do Tell Tales': teasing out the contribution of the folk hero to heritage-based tourism. [Doctoral thesis. La Trobe University, Melbourne].
Fagence, M. (2023). Enrichment and enlightenment from engagement with history through heritage-based tourism. In T. V. Singh, R. Butler, & D. Fennell (Eds.), *Tourism as a pathway to hope and happiness* (pp. 152–169). Channel View Publications.
Fagence, M. (forthcoming). 'Walking battlefield sites': An experience of secular pilgrimage. In V., Ziakas, C., Lundberg, & M. Lexhagen (Eds.). *Popular culture destinations, fan pilgrimages and touristic world-making*. (pp. 53–74). Channel View.
Filep, S., & Pearce, P. (2014). *Tourist experience and fulfilment*. Routledge.
Fryer, T. (2020). Reflections on positionality: Archaeological heritage praxis in Quintana Roo, Mexico. *Archaeological Papers of the American Anthropological Association*, *31*, 26–40.
Hoggan, C., & Finnegan, F. (2023). Transformative learning theory: Where we are after 45 years. *New Directions for Adult and Continuing Education*, *177*, 5–11.
Hosany, S., Sthapit, E., & Bjork, P. (2021). Memorable tourism experience: A review and research agenda. *Psychology and Marketing*, *39*, 1467–1496.
Hosseini, S., Macias, R., & Garcia, F. (2021). Memorable tourism experience research: A systematic review of the literature. *Tourism Recreation Research*, *48*(3), 465–479.
Hosseini, E. Z., Hashemi, M., & Hajikazemi, S. (2023). Sustainable Regionalism in Architecture: A Framework of Principles and Components. *Prometheus*, *6*, 38–43.
Hyde, K., & Harman, S. (2011). Motives for a secular pilgrimage to the Gallipoli battlefields. *Tourism Management*, *32*, 1343–1351.
Kerstetter, D., & Cho, M.-H. (2004). Prior knowledge, credibility and information search. *Annals of Tourism Research*, *31*(4), 961–985.
Kim, J.-H., Ritchie, J., & McCormick, B. (2012). Development of a scale to measure memorable tourism experience. *Journal of Tourism Research*, *51*(1), 12–25.
Kirillova, K., Lehto, X., & Cai, L. (2016). Tourism and existential transformation: An empirical investigation. *Journal of Travel Research*, *56*(5), 638–650.
Landsberg, A. (2004). *Prosthetic memory: The transformation of American remembrance in the age of mass culture*. Columbia University Press.
Lennon, J., & Foley, M. (2000). *Dark tourism: The attraction of death and disaster*. Continuum.
Liu, Y., Dupre, K., & Jin, X. (2021). A systematic review of literature on contested heritage. *Current Issues in Tourism*, *24*(4), 442–465.
Lloyd, D. (1998). *Battlefield tourism: Pilgrimage and the commemoration of the Great War in Britain, Australia and Canada, 1919–1939*. Bloomsbury.
Logan, W., & Reeves, K. (2009). *Places of pain and shame: Dealing with difficult heritage*. Routledge.
Lundberg, C., & Ziakas, V. (Eds.). (2019). *The Routledge handbook of popular culture and tourism*. Routledge.
Macdonald, S. (2009). *Difficult heritage*. Routledge.
McQuilton, J. (1979). *The Kelly outbreak: The geographical dimensions of social banditry*. Melbourne University Press.
Mezirow, J. (1991). *Transformative dimensions of adult learning*. Jossey-Bass.

Miles, S. (2016). *The Western front: Landscape, tourism and heritage*. Pen and Sword Books.

Mitchell, J. (2006). Case and situational analysis. In M. David (Ed.), *Case study research* (vol. 2, pp. 59–83). SAGE.

Moustakas, C. (1990). *Heuristic research: Design, methodology and application*. SAGE.

Nandasena, R., Morrison, A., & Coca-Stefaniak, J. (2022). Transformational tourism – A systematic literature review and research agenda. *Journal of Tourism Futures*, *8*(7), 1–16.

Nawjin, J., & Filep, S. (2016). Future directions for future tourists well-being research. *Annals of Tourism Research*, *61.C*, 221–223.

Palmer, C. (2009). Reflections on the practice of ethnography within heritage tourism. In M. Sorensen, & J. Carman (Eds.), *Heritage studies: Methods and approaches* (pp. 123–139). Routledge.

Palmer, C., & Tivers, J. (Eds.). (2019). *Creating heritage for tourism*. Routledge.

Pung, J., & Chiappa, G. (2020). An exploratory and qualitative study on the meaning of transformative tourism and its facilitators and inhibitors. *European Journal of Tourism Research*, *24*, 1–21.

Pyke, S., Hartwell, H., Blake, A., & Hemingway, A. (2016). Exploring well-being as a tourism product resource. *Tourism Management*, *55*, 94–105.

Robinson, P. (2015). I remember it well: Epiphanies, nostalgia, and urban exploration as mediators of tourists memory. *Tourism, Culture and Communication*, *15*, 87–101.

Rus, K., Dezsi, S., Ciascai, O., & Pop, F. (2022). Calibrating evolution of transformative tourism. *Sustainability*, *14*(7), 1–40.

Ryan, C. (Ed.). (2007). *Battlefield tourism: History, place, and interpretation*. Elsevier.

Scates, B. (2016). *Return to Gallipoli: History, place and interpretation*. Elsevier.

Scott, N., & Le, D. (2017). Tourism experience: A review. In N. Scott, J. Gao, & J. Ma (Eds.), *Visitor experience design* (pp. 30–49). CABI.

Seligman, M. (2002). *Authentic happiness*. Free Press.

Seyfi, S., Hall, C., & Rasoolomanesh, S. (2019). Exploring memorable tourism experiences. *Journal of Heritage Tourism*, *15*(3), 341–357.

Sheldon, P. (2020). Designing tourism experiences for inner transformation. *Annals of Tourism Research*, 83, Article 102935.

Silverman, H. (2011). Contested cultural heritage: A selective historiography. In H. Silverman (Ed.), *Contested cultural heritage* (pp. 1–49). Springer.

Sirgy, M., & Uysal, M. (2016). Developing a eudaimonia research agenda in travel and tourism. In J. Vitterso (Ed.), *Handbook of eudaimonic well-being* (pp. 485–495). Springer.

Sobchack, V. (1996). *The persistence of history*. Routledge.

Soulard, G., McGehee, N., & Stern, M. (2019). Transformative tourism organizations and globalization. *Annals of Tourism Research*, *76*, 97–104.

Sthapit, E., & Coudounaris, D. (2017). Memorable tourism experiences: Antecedents and outcomes. *Scandinavian Journal of Hospitality and Tourism*, *18*(1), 72–94.

Strijbosch, W. (2022). Experience Unpacked. [Doctoral dissertation, Tilburg University].

Taylor, E., & Cranton, P. (2013). A theory in progress? Issues in transformative learning theory. *European Journal for Research on the Education and Learning of Adults*, *4*(3), 33–47.

Teo, M., Wang, Y., & Kwek, A. (2021). Conceptualizing co-created transformative experiences a systematic narrative review. *Journal of Hospitality and Tourism Management*, *47*, 176–189.

Tunbridge, J., & Ashworth, G. (1996). *Dissonant heritage: The management of the past as a resource in conflict*. Wiley.
Uzzell, D., & Ballantyne, R. (1998). Heritage that hurts: Interpretation in a postmodern world. In D. Uzzell, & E. Ballantyne (Eds.), *Contemporary issues in heritage and environmental interpretation* (pp. 152–171). The Stationary Office.
Voigt, C., Brown, G., & Howat, G. (2011). Wellness tourists: In search of transformation. *Tourism Review*, *66*(1–2), 16–30.
Wall, S. (2016). Toward a moderate autoethnography. *International Journal of Qualitative Methods*, *15*(1), 9pp.
Yin, R. (2003). *Case study research: Design and methods* (3rd ed.). SAGE.
Yin, R. (2018). *Case study research and applications: Design and methods* (6th ed.). SAGE.
Zhao, Y., & Agyelwaah, E. (2023). Understanding tourists' transformative experience: A systematic literature review. *Journal of Hospitality and Tourism Management*, *54*, 188–199.

12

TRANSFORMATIVE SUSTAINABLE MARKETING DOCTRINE FOR HERITAGE TOURISM

Deepak Chhabra

The contributing authors focus on a broad spectrum of areas/case studies from across the globe and focus on topics relevant to most of the sustainable development goals (SDGs). They identify issues impeding the sustainable planning and management of heritage tourism and suggest innovative solutions. Jolliffe et al. (2024) reiterate that heritage tourism needs to be developed, implemented and marketed in a sustainable manner, to generate a transformative impact. The authors use tea business enterprises, in Thailand, as a case in point and illustrate that experiential aspects of tea traditions hold tremendous potential to generate economic benefits and transform both the guests and the hosts. The authors emphasize that host community involvement is a critical aspect of the objectively authentic tea tourism experience. Taking a supply perspective, Viana-Lora (2024) stresses on the significance of sustainable marketing of heritage tourism to advocate long-term preservation of local heritage. The author highlights the significance of strategic planning to produce long-term sustainable benefits for the current and future generations. Sawhney et al. (2024) focus on tribal tourism in India and identify issues associated with their rights and well-being. The authors suggest a regulatory approach to ensure empowerment of the tribes by taking their views and cultural values into consideration and shaping the manner in which their heritage is promoted. They note that concerns have arisen regarding the exploitation of indigenous communities and the neglect of their rights and well-being. It is suggested that the ultimate aim of promoting tribal tourism should be to augment indigenous empowerment, preservation and viable economic benefits for the tribes.

Fagence (2024) employs a case study approach and focuses on a specific story belonging to a piece of Australian history from a chaotic time period of

DOI: 10.4324/9781003346197-12

time. He takes a unique initiative to illustrate that transformative experiences can follow a complex path as they might be subjected to several viewpoints and dissonant interpretations. He argues that conflicting interpretations can pose a dilemma to the stakeholders who are committed to tell the story in an authentic manner. Furthermore, tourists identify with the version that is most palatable for them. Xie (2024) focuses on perceptions of residents with regard to the staycation programs launched by the government of Macau as part of its effort to revitalize tourism. These programs are the outcome of the pandemic times and are delineated into routes based on heritage and nature-based resources. The author reports that the staycation programs contribute to the well-being of the guests in terms of peace of mind, connectedness and happiness.

Timothy (2024) focuses on the emerging popularity of cultural routes and examines multiple benefits of cultural heritage in the form of the "Cultural Routes" program in Europe. Such initiatives inspire personal growth and regional collaboration and harmony. The chapter takes a discursive view of the role of heritage tourism in revitalizing regions from catastrophic events both natural (such as pandemic) and man-made (such as wars) by boosting economy, solidarity and overall well-being. The core aspects of the proposed sustainable marketing model in Chapter 1 (partnership, collaborations, and personal growth/well-being) are supported in Timothy's chapter. Yilmazdogan and Ece (2024) in their chapter on retrobranding discuss the manner in which historical, cultural, or natural heritage sites can utilize strategic marketing strategies to stimulate appeal, integrity preservation, and tourist engagement. The authors scrutinize several case studies to detect pros and cons of rebranding and retrobranding to help stakeholders plan strategies grounded in sustainable heritage management and marketing. They reveal that retrobranding can serve as an effective marketing tool to promote different aspects of sustainability. He (2024) reveals that objective authenticity is not necessarily the sole predictor of satisfaction of heritage tourism experience. Based on content analysis of tourist comments on local homestay websites, she reports that tourists seeking authenticity also give consideration to physical comfort. Tourists can develop a negative attitude toward the tangible environment even if it denotes the purest version of authenticity. As an instance, accommodation experience can shape the satisfaction level of an objectively authentic location. In other words, degree of physical comfort and service at a heritage location can influence satisfaction levels. This implies that some heritage tourist segments are not likely to value disoriented experiences (which can take them beyond their comfort zones), although they might be objectively authentic (that is, offer an expression of the most pristine authentic version).

Several studies note that objectively authentic experiences can be uncomfortable because they are not modified to suit comfort needs of some tourists such as those belonging to the psychocentric category. Nevertheless, there are other types of tourists such as allocentrics and serious leisure travelers who might be backpackers, adventure tourists and/or volunteer tourists with preference for disorienting experiences that take them beyond their comfort level (Tiberghien et al., 2020). It can be argued that the disoriented experiences hold the most potential to trigger optimal flow and transformation of the inner self (Tiberghien et al., 2020). From the standpoint of transformation, as He's tourists desire a hedonic experience, hosts might feel encouraged to modify some of their objectively authentic offerings. In such cases, guided sustainable marketing initiatives can be taken to inform both the guests and the hosts to pursue a viable sustainable path. Pintel and Chhabra's (2024) chapter discusses the pros and cons of employing a popular visitor engagement technique by museums (folksonomy and social tagging). The authors point out that museums have to evolve to sustain interest of their audience. Therefore, they are called upon to develop innovative programs to capture the attention of their patrons. Their study explores Gen. Z's perceptions of folksonomy and social tagging to determine if this process is effective in captivating the attention of the younger generation. Furthermore, they seek views about the responsible employment of this tool. Pintel and Chhabra (2024) report that the majority of the Generation Z respondents feel that folksonomy and social tagging augmented their curiosity about the exhibits and enriched their knowledge. Most of the generation Z cohort also felt that including folksonomic terms in museum records is not an abrogation of museum responsibility. Instead, this tool sparks conversation and offers additional insights. Furthermore, it motivates the audience to seek actual information on the exhibit to match with their tags. A few, however, expressed concerns and stated that the museums should ensure that the techniques they employ are not compromising their core responsibility toward preservation and delivery of an accurate interpretation of exhibits. The authors suggest a strategic participatory approach grounded in citizen science and citizen humanities frameworks. It is further recommended that the museums should engage the public with exhibits that convey meaningful messages associated with sustainability of heritage and well-being of society, to mindfully support some of the SDGs proposed by the United Nations.

In a nutshell, the authors of this book reiterate the impact of mindful, engaging and strategically planned initiatives on conservation and preservation of heritage resources and in safeguarding the interests of the host community. For the most part, the different elements of the sustainable marketing model (proposed in Chapter 1) are supported. Several sub-sets of heritage tourism are also highlighted that hold potential to promote sustainable use of resources and well-being of guests and hosts, particularly staycation programs, traditional homestays, cultural routes, and community-based tea tourism. Social tagging of museum exhibits can facilitate learning and engagement with the museum exhibits. Sustainable benefits include revitalization in the face of unforeseen calamities, preservation ethos, local empowerment, and stakeholder collaboration. Wellness related benefits include engagement/immersion, personal growth, and mindfulness. To cement the positive impacts of the aforementioned activities and sustain the state of well-being, it is important to elevate them to a transformation stage and this stage requires regulatory interventions and commitment from the suppliers, hosts and the tourists. As pointed out by Fagence (2024), the notion of transformative heritage tourism itself can trigger multiple and dissonant views. Therefore, the transformative process needs to be inclusive of local voices and involve a broad spectrum of heritage tourism stakeholders. Regulations are also needed to ensure equitable use and distribution of resources and tourism-related benefits, for instance, for the tribal communities. Moving forward, based on a cursory review of recent literature, the next section offers insights on the prevailing topics in recent literature on heritage tourism. It is followed by an overview of contemporary trends from a practitioner viewpoint.

Emerging Themes in Scholarship

A perusal of documented literature, in the last decade, reveals focus on a variety of notable topics such as authenticity, memoryscapes, multi-sensory experiences, psychological aspects of experience (for instance, nostalgic emotions), intangible heritagization, heritage engagement paradigms, host community attitudes and involvement, and innovative methods to stimulate cross-sector and regional collaborations. Emerging popularity and potential of heritage trails and heritage routes/corridors as engaging and sustainable spaces is noted.

For instance, Duda (2023) examines the Griffins' Route which is a new transboundary tourist project in Europe. Its theme is based on shared history and cultural heritage of the Duchy of Pomerania. The author notes that differences in the manner the visitors perceive the route and its amenities and infrastructure are based on their country of birth, emotional engagement, and trip motives. The authors report that such destinations are interpreted base on different priorities and political agenda. Emotions and emotional responses also constitute an important area of study in heritage tourism (Prayag & Del Chiappa, 2023; Rasoolinamesh & Lu, 2024). For instance, Rasoolinamesh and Lu (2023) use SOR (stimulus-organism-response) theory to determine the intervening influence of memorable tourism experiences (MTE). Psychological aspects of experience, that can stimulate nostalgic emotions, are highlighted by several studies (Earl & Hall, 2023; Prayag & Del Chiappa, 2023). Thirumaran et al. (2023) also study the emotional aspects of heritage experiences and evaluate the role of motivation, tourism impacts, emotional access, and sentiments in generating tourist satisfaction at heritage sites. Several studies report significant association between host community participation in heritage tourism-related decisions and their sustainable livelihood (Chen & Wan, 2023) and point out the poor management of resources by stakeholders with conflicting interests can compromise the quality of life of local residents and quality of experience of visitors (Vidal-Matzanke et al., 2023).

Additionally, the pandemic times served as a wakeup call for heritage institutions to recognize the significance of technology. The pandemic closed the doors of many heritage attractions and those which remain active were the ones who connected with their audience through social media platforms and several other online networking tools. In recent studies, several scholars have illustrated the significance of integrating technological innovations in heritage tourism (Nikolaou, 2024). Intangible heritagization has emerged as another focal area of inquiry. As an instance, del Mármol and Santamarina (2023) offer a discourse on the manner in which intangible heritage policies have become a crucial component of city branding strategies as it fosters local identity and sense of place. Furthermore, some recent studies, particularly in the post-pandemic times, have scrutinized the potential of heritage tourism to boost the well-being of the visited and visiting communities. Heritage engagement paradigms have been suggested to overcome barriers to immersive interactions. Literature continues to focus on world heritage sites and several authors have shared the political pitfalls of the designations which ignore the voices of the locals. For instance, Hashimoto and Telfer (2024) share that "outsider interference, distortion of historical facts, staging and the Disneyfication of heritage are contentions raised over how applications are manipulated to achieve a UNESCO recognition. Greater recognition of local voices and alternative narratives is critical in the designation process to be true to history in an age dominated by heritage" (2024, p. 221).

Furthermore, the nature and intensity of host and guest interactions, based on selection of natural or man-made space settings, continues to remain a crucial line of inquiry and several studies have taken the significance to a higher level to scrutinize sustainable and transformative potential of such connections. For instance, Nguyen et al. (2024) examined the host-visitor interactions in the Central Highlands of Vietnam based on the CMM (Coordinated Management of Meaning) theory and noted that most interactions happened in private spaces (such as homes), tourist attractions and facilities and on tours. Based on the nature of the setting, depth of connectivity varied. Barriers to meaningful interactions are noted to include verbal and non-verbal behavior and cultural patterns. Prospects to facilitate favorable interactions in ethnic tourism settings are shared. Tourist-to-tourist interactions can also influence positive or negative experiences at a heritage destination and overall satisfaction levels. Trupp et al. (2023) suggest management strategies to stimulate optimal experiences of such interactions.

Paranormal investigations on YouTube, an emerging type of armchair travel is becoming popular. It is also termed as virtual dark tourism and has helped unveil familiar dark tourism places. Basaraba (2024) outlines pitfalls, hazards, and ethical concerns for tour guides and researchers seeking this type of tourism in the future. Techniques applied are digital ethnography and distant and close reading of a broad spectrum of YouTube channels focusing on ghost hunting. Furthermore, heritage tourism has helped revive interest in declining industries and spaces. For instance, Szromek and Butler (2024) report that tourist visits in declining post-industrial areas have justified the need for investment in such areas. The authors scrutinize "the present re-development of tourist attractions in Upper Silesia area (in Poland) based on features of the industrial heritage there, including both movable monuments .... as well as fixed features such as ......, a historic worker's housing estate and a wooden radio station. They suggested refocusing of the IMR (Industrial Monuments Route) is part of the process of conserving the endangered regional identity for future generations" (Szromek & Butler, 2024, p. 320). Drawing on poor rural communities in a developing country (Nepal), Pradhan et al. (2022) illustrate the manner in which cultural tourism can assist in offering a passage to develop host communities and eliminate poverty. The core message imparted by this study is aligned with the work of several contributors of this book (such as Jolliffe et al., 2024; Timothy, 2024). Pradhan et al. (2022) contend that active participation of the local community is essential and it is useful to engage outside organizations in the process to ensure equitable distribution of benefits and cultural and environment integrity. A community-based entrepreneurship model was used. Lund et al. (2022) propose an experience strategy framework to develop novel experiential activities and programs. Trupp et al. (2023) report unequal benefits and sociocultural aspects of handicrafts and souvenirs from the viewpoint of female vendors and microbusinesses. The economic benefits are mainly accrued from cruise-ship tourism and result in unequitable distribution of revenue.

Some studies have examined resident perceptions and support for heritage tourism based on social exchange theory (Allaberganov & Catterall, 2023). According to Dragin et al. (2023), resident perceptions of sustainable tourism development are shaped by their socio-demographic characteristics. Authenticity continues capture attention particularly in the context of its social value and contribution to health and well-being. Based on the foregoing, it is important to integrate a multitude of topics into the heritage planning and marketing agenda such as authenticity, promotion of local identity and place attachment, emotions such as nostalgia, conservation/preservation, branding, digital innovation, atmospherics, and experiential dimensions of heritage resources (Chhabra, 2017).

Emerging Trends

Next, numerous emerging trends in tourism (that are of relevance to heritage tourism) are identified so that appropriate marketing messages, to promote sustainable and transformative heritage tourism, can designed for target markets. These include:

- *Acting on sustainability*: Travelers are seeking quiet and uncrowded destinations to decrease overtourism. They are planning their travel in a conscious and mindful manner and are willing to pay more for environment-friendly and culturally mindful programs;
- *Immersing in culinary travel*: Culinary travel is an emerging trend where travelers desire to learn and immerse themselves in other cultures through localized and traditional culinary experiences. Travelers desire to attend privately hosted dinners, patronize open-air food markets and friendly interactions with local residents;
- *Wellness travel is a priority*: Demand for wellness travel has surged. Self-care has become the priority for most citizens across the globe. More and more Americans are expressing their desire to travel to "relax, recharge, to disconnect and to prioritize their health";
- *Engaging travel shaping choices*: Preferences are noted toward engaging activities such as biking, hiking, wildlife and cultural experiences that offer a sense of restoration, self-fulfillment, connectedness and a sense of achievement. In other words, demand is for experiences that contribute toward overall well-being;
- *Demand for more shoulder season gateways*: The pandemic led to a surge in digital nomads and other types of remote workers. Because of the flexibility of being able to work away from office, people now have the autonomy to travel throughout the year. Particularly, shoulder seasons offering solitude and lower rates, are being sought
- *Longer itineraries*: Along the lines of revenge travel (taking back your time because of the halt in travel triggered by the pandemic), travelers are interested in staying longer and seek immersive experiences;
- *Passion first, destination second*: Travelers today prefer to travel to a destination they are passionate about, that is, a place that is of interest to them and offers their favorite activities and experiences;
- *Set-jetting*: Preferences are being inspired and shaped by places shown in movies and TV shows;
- *Slow-life travel*: Consumers prefer slow paced and multi-sensory experiences to help them relax and recuperate from the daily busy life and;
- *Familiar places in unfamiliar ways*: Preferences are soaring for customized and personalized experiences to familiar places. The guests can have unique experiences in familiar places such as by substituting trails or

immersing themselves in a staycation experience (as shared by Xie, 2024), museum exhibit or taking a guided tour. Such ways of experiencing, the familiar from a different way, can give a fresh view of the destination, heritage and sceneries (Altschuler, 2024; Taylor, 2024).

Based on the foregoing, nostalgia continues to be scrutinized by heritage tourism scholars (Adie & de Bernardi, 2023). Other topics include sensory experiences, resident perceptions, seeking personal attachments through diaspora travel, emotions, emotional accessibility, and emotional intelligence. Because authenticity and digitalization continue to receive unprecedented attention by both scholars and practitioners, the following paragraphs scrutinize them in an in-depth manner from a sustainable marketing and transformation standpoints.

Authenticity

Authenticity is notably being scrutinized, in the recent times, as a conduit for self-improvement and for cultivating a desire to use and promote sustainable use of heritage resources and in stimulating the well-being of the guests as well as the host communities. Factors influencing consumer perceptions of authenticity also remain a popular line of inquiry. Some of the novel perspectives of authenticity, based on a review of recent literature, are discussed in the next few paragraphs: balanced authenticity, the Eastern dialectical theory, authenticity and well-being, and authenticity from the viewpoint of the hosts. In context of the host community, the notions of localized identity, cultural self-improvement, and heritage responsibility are touched upon.

Notion of Balanced Authenticity

From an existentialist standpoint, some scholars have described authenticity as a way of being that mirrors the inner self through an honest portrayal of one's emotions, thoughts and feelings. However, it is argued that the true self can be impacted by external factors and therefore the actual portrayal of the self might undergo alteration to connect with the external surroundings. Human beings are social beings and their way of living and portraying themselves is shaped by the social environment around them (Wang, 2016). Chen et al. (2006) argue that the true self simultaneously exists with the social (or relational) self. According to Wang, "the most challenge is not to realize one's inner thoughts through eliminating others' influences, but rather how to obtain one's goal in the environmental restrictions and then to achieve harmonious unity between true self and relational self" (2016, p. 316).

Wang presents three notions of authenticity: egocentric authenticity, other-distorted authenticity, and balanced authenticity. The notion of egocentric authenticity refers to poor satisfaction, from a relational perspective, and power exerted by dominant authorities. Other-distorted authenticity denotes uncontrolled communion and low autonomy. Balanced authenticity is shown to predict both agency and communion, and is noted to have a positive correlation with subjective well-being (SWB), even while controlling the social desirability factor. There is an ongoing conflict in daily lives between the intrinsic/true self and the self that is presented to the social world which is shaped by personal needs and demands of the external environment. Two responses can result when human beings encounter external forces: either they can be true to themselves and speak out their mind or they can conceal their inner thoughts. The former is regarded as an authentic response while the latter refers to inauthenticity. Some scholars posit that the fear of being socially shunned results in inauthentic responses and this fear can negatively impact individual well-being (Neff & Harter, 2002; Sheldon et al., 1997). Furthermore, the extreme position where authenticity is exemplified by lack of consideration for the perspectives of others (termed as egocentric authenticity) (Schmid, 2005), can result in discordant social connections and lower levels of well-being (Chen et al., 2001; Kwan et al., 1997). In a nutshell, Wang's (2016) work offers insights on the relational crux of authenticity and its consequent connection with well-being.

Based on the foregoing, to address the shortcomings of egocentric authenticity and inauthenticity, Wang (2016) presents the notion of balanced authenticity premised on balancing the intrinsic self with the external factors that are likely to influence a person. He argues that both agency and solidarity (communion) are essential for optimal SWB and psychological well-being (Keyes et al., 2002). Theories that have offered insights on ideal ways to strike a harmonious compromise between the true inner self and camaraderie include the optimal distinctiveness theory *(*which stresses on striving for an optimal balance of inclusion and distinction in an inter-related and intra-related manner with social groups and events*)* (Brewer, 1991), functional flexibility theory (referring to the hybrid role of employees tasked with different roles and responsibilities (Paulhus & Martin, 1988) and Eastern dialectical thinking (Peng & Nisbett, 1999).

The Eastern Dialectical Theory

The Eastern dialectical theory stresses on the significance of reciprocal consideration of both views to ensure optimal outcomes. From this standpoint, "authenticity is a process term that encompasses unity and plurality, which ultimately contributes to the 'fully functioning person' (Schmid, 2001). When confronted with a conflict between agency and communion, highly dialectical individuals will neither stand by their own ideas to maximize self-interests, nor easily abandon their own opinions................ Instead, they will maintain a balance between internal and external pressure and find a solution that will be broadly accepted" (Wang, 2016, p. 317). Grounded in the Eastern dialectical theory, Wang conducts an empirical scrutiny of the manner in which balanced authenticity optimizes subjective and psychological well-being. His study also tests whether balanced authenticity can predict optimal well-being and simultaneously promote solidarity. He develops AIRS (authenticity in relationship scale) underpinned on three concepts of authenticity (oriented by self, oriented by others and balanced).

Authenticity and Well-being

Wang reports significant correlation between balanced authenticity and SWB and psychological well-being. These findings reiterate that authenticity plays a key role in promoting overall well-being (Kifer et al., 2013; Wood et al., 2008). Wang furthers this line of inquiry by delineating authenticity and reporting that the balanced version of authenticity contributes to well-being and therefore balanced authenticity plays a superior role in envisaging well-being. This stance aligns with the negotiated and theoplacity strand of authenticity reported by several scholars. In fact, it extends the implications of the negotiated theories from a well-being standpoint.

Host Perspectives of Authenticity

Furthering the impact of authenticity from the host community viewpoint also constitutes an important line of inquiry. In fact, the fundamental tenet in sustainable development of cultural heritage resources is eliciting views of the host community. For instance, perceived authenticity can generate an impact on the host identities and trigger cultural self-improvement and a sense of cultural responsibility toward the development of cultural heritage in their area (Yi et al., 2024). The authors reports that "local residents' localized identity and cultural self-improvement, activated by perceived authenticity of the cultural heritage, significantly stimulate their heritage responsibility" (Yi et al., 2024, p. 1). Also, tourist perceptions of cultural heritage and the nature of their authentic experience is shaped by the manner in which the hosts conserve and present the core attributes of their heritage such as physical forms, cultural and historical connections, and active transmission of meanings. These core elements can shape tourists' perceived authenticity of cultural heritage and their authentic experience (Farrelly et al., 2019). Along similar lines, however, "loss of authenticity and originality of cultural heritages can significantly affect heritage management and local residents' attitudes and behaviors," in regard to their involvement in community initiatives and advocacy of sustainable tourism (Jaafar et al., 2015; Yi et al., 2024, p. 1).

Limited studies have offered insights on how locals perceive the authenticity of their cultural heritage. Heritage experiences by locals foster personal, shared or national identity (Butler et al., 2014). Yi et al. study a sense of local identity from an eastern ideological perspective. As pointed out by Tuan (1974), truth does not derive solely from an objective assessment of elicited information; it is also derived from experiences at the personal level that result in subjective acknowledgment and adoption of authenticity. Tuan argues that personal experience is critical because of the animate nature of the human body and the environment or space is "humanly understood or construed" (1977, p. 35). That is, it is through their personal interaction with space or the surroundings around them that human beings are able to experience, feel, and comprehend it.

Localized Identity

From the foregoing, it can be deduced that localized identity does not only refer to cultural identity but it has historical, social, and interactive connotations. From the standpoint of Breakwell's (1986) identity process theory, localized identity is the outcome of an interactive process between people and the environment where they reside, and this "involves social classifications, structure, consumption modes as well as traditional inherited customs and habits. In the heritage setting, localized identity denotes how local residents' identities culturally and socially link to the cultural heritage sites where they live (Hunter, 2011; Zhou et al., 2015). History and traditions, ... cultural activities and events (e.g., local cultural festivals) are proven to help local people to recognize their unique history and strengthen their identities" (Yi et al., 2024, p. 3). Timothy's (2024) chapter on cultural routes reiterates this point, that is, heritage can be served as a conduit to promote a sense of rootedness, self-fulfillment and can assist in the recovery of communities devastated by a crisis.

Cultural Self-improvement

Cultural self-improvement implies that a person's cultural attributes evolve over a period of time (Tan et al., 2018). This notion refers to two components:

1 "self" signifying one's own identity that relies on his/her strength, highlights his/her characteristics, and takes his/her cultural path;
2 "improvement" depicts the strong influence, vitality, creativity, and competitiveness of the cultural advancement.

(Yun, 2010)

It is posited that awareness of one's weaknesses and drawbacks are essential for cultural self-improvement and positive actions need to be implemented to ensure cultural continuity. The entire process engages with the manner a person upholds, safeguards, and cultivate his or her culture, such as cultural autonomy (Yun, 2010). Local culture therefore evolves with time and its re-definition assists in adapting to the changing times particularly from a social standpoint (Geertz, 2017; Tuan, 1974). In a nutshell, cultural integration and change bestow culture with a positive strength described by four attributes:

1 culture can self-recover or self-repair when it is broken or damaged;
2 culture is changeable and dynamic, which needs to be created under established conditions inherited from the past;
3 culture can be extended while maintaining itself; and
4 culture has its own driving force or power.

(Geertz, 2017)

Heritage Responsibility

Yi et al. refer to heritage responsibility as "accountability that an individual undertakes in the process of cognition, interpretation, and reproduction of heritage value" (2024, p. 3) while taking legal, moral, mental, economic, and social responsibilities into consideration (Zhang, 2014). From a broader standpoint, it can be regarded as social responsibility of an individual or organization. A review of documented literature shows that responsible dispositions of tourists have received much attention in research albeit little thought has been devoted to the heritage responsibility of local residents. The locals embrace dual roles as host anchors and as experiencers of cultural heritage and can be admirers and fortifiers at the same time (Zhou et al., 2015). The local community voices play an important role in understanding and identifying ways to protect cultural heritage (Tan et al., 2018). This point is reiterated by Sawhney et al. (2024) in Chapter 7. Yi et al. speak that "when local residents realize that their cultural traditions are vanishing, they are likely to take action to preserve cultural heritage" (2024, p. 3). By developing a heritage responsibility scale for both the tourists and residents, Li et al. (2019) share that heritage responsibility can be expressed through a sense of responsibility for the self and for others. Examples include protecting heritage, cultivating awareness, communicating and informing others and influencing other people's disposition toward heritage.

Mediating Sustainable Behavior

Some studies note that place attachment and cultural motivations have a positive impact on commitment toward preservation, environment and sustainable tourism (Albrecht & Raymod, 2021; Cordina et al., 2022; He et al., 2021). The important task is to identify tools, platforms and/or mediators that can trigger and reinforce this commitment such as interpretation, experiential quality, message framing and touching.

Interpretation: Wang et al. (2024) suggest that interpretation can offer a useful opportunity to cultivate and guide the cultural preservation commitment of tourists; as it disseminates information related to culture and heritage through a myriad of ways and synchronizes experiential, cultural preservation and sustainable management goals (Huang et al., 2022). Multiple interpretation techniques are employed by heritage institutions around the globe albeit several focus merely on entertaining and imparting factual information (Timothy, 2021). Wang et al. present the following tools:

- Use of stories, metaphors, and other techniques to develop storylines and interesting themed interpretation forms such as historical story books and themed comics;
- Use of vivid and educational content to explain historical heritage and celebrity stories to inspire tourists' affective resonance and preservation commitment; and
- Describing heritage stories using virtual characters, integrating cultural features into character sentiment, and organizing episodic and anthropomorphic interpretation activities also help guide tourists' preservation commitment.

(2024, p. 17)

Experiential quality (EQ): Several studies have shown that quality of experience holds tremendous potential to shape tourist intentions and anticipate their level of satisfaction (Moon & Han, 2019). Because it has service-related connotations, EQ refers to experience associated with the tangible environment, interactive levels and heritage attributes (Hussein & Hapsari, 2021). Most heritage tourism destinations make use of interpretive brochures and tourism websites. However, these offer limited inspiration to tourists because they fall into the category of factual content (basic facts about the destination). Quality of experience is more influenced by emotional content and stimulation of all senses, sentiments and activities. According to Moon and Han (2019), tourists desire for deep experiences engaging all senses. Wang et al. write that "heritage tourism interpretation, as a tool of information display and heritage experience, fosters intimate interaction and connection between tourists and heritage destination culture, inspiring new discoveries, ... and insights" (2024, p. 6). Based on the narrative transportation theory, emotional narratives easily draw tourist intention as they move them into the narrative realm and affective engagement augments memory and acknowledgment of the narrative and stimulation of empathy for the heritage resource (Ryu et al., 2019) leading to a better-quality experience.

Message framing: Some research shows that the manner in which a message is framed shapes the personal behavior of tourists (Chi et al., 2021; Zhang et al., 2018). Framing can employ a loss or gain format. As an instance, Chang et al. (2015) gain (vs. loss) framing influences the purchasing intentions toward green products. Along similar lines, Grazzini et al. (2018) notes that guests of a hotel are more likely to embrace recycling activities if they are exposed to loss framing, Thus, message framing, as a key way to interpret content information, may influence the relationship between interpretation content and persuade commitment toward preservation (Zhang et al., 2018).

Touching: Quality of experience is more enhanced by touching (affective stimuli) than cognition, (that is, factual content). In other words, interpretation on affective content holds a more significant transportation function and enhances a sense of immersion and touch (Cordina et al., 2022). Wang et al. (2024) report that particularly in the context of heritage tourism, "tourists seek intimate experiences and affective connections with cultural heritage (Santa & Tiatco, 2019). Accordingly, affective (vs. factual) content allows tourists to embrace cultural implications in interpretation and strengthens their affective engagement and active commitment" (2024, p. 15).

Next point for deliberation focuses on a tool that can stimulate deep-rooted transformation in thinking and behavior and emotionally rich engaging experiences that can result in memorable moments and reminiscence. As mentioned in Chapter 1, Metaverse is an emerging trend and holds potential to effectively engage tourists. However, it is also subject to numerous challenges and limitations. The next section delves deeper into the challenges posted by this innovative platform.

Challenges of Metaverse

Although Metaverse is the new emerging technological trend with potential to integrate actual and virtual spaces, it also poses several challenges that compromise value and fog reality by causing confusion between what is real and what is virtual. This can result in diminished critical thinking and creativity impeding the capability of users to deliberate on their experiences (Qin et al., 2022). The challenges and threats associated with Metaverse can be delineated into technological, economic, experiential and ethical/legal categories:

- *Technological*: Metaverse is at a conceptual stage and its platforms have not been fully developed. Therefore, best practice examples or a blueprint do not exist, to ensure its smooth functioning. Organizations are still reluctant to venture into untested waters. The entire process is risky and expensive and calls for innovation and creative imagination and financial investment;

- *Economic*: Tourism destinations and enterprises across the globe do not carry sufficient economic and technological resources for investment. Because it uses (equipment such as sensory clothing and virtual reality and Metaverse headsets are expensive) cutting-edge platforms which are costly, it is likely to privilege a few high-end organizations and marginalize tangential businesses and poor regions. Initial investment is required and high expenses can lead to inequitable growth as disadvantaged regions will be left out;
- *Experiential*: According to Pratt (2022), digital immersion can lead to motion sickness for those who are able to afford it. Engaging with immersive experience can be challenging for some groups of users;
- *Ethical and legal*: Virtual environments also pose ethical and legal challenges as they can become havens for "illegal and immoral behaviour and inappropriate legislation and policing. This is particularly complex in tourism since users make cross borders and applicable laws may be location-based. In virtual environments, people hide behind their virtual personas and adopt" irresponsible, bullying and other aggressive and harassing behaviors such as cyberstalking, fraudulent activities, money laundering, and drug abuse. Virtual policing can pose a different set of law enforcement challenges (Qin et al., 2022);
- *Addictive behavior*: As mentioned earlier, Metaverse can blur the boundary between the physical and virtual environments and offer illusive experiences and trigger addictive behavior for those who are likely to use it to escape from their problems. Furthermore, overindulgence in digital activities such as virtual games can have a negative impact on the physical and psychological health
- *Intellectual property and copyright issues*: They need to be addressed which ensures that tourism brands and destinations are sheltered. Metaverse is also subject to security and privacy-related issues (Buhalis et al., 2023). Furthermore, liability of privacy and security issues and their handling agency are areas of concern particularly if multiple organizations and business are engaged in designing and influencing the Metaverse experiences (Buhalis et al., 2023; Pew Research Center 2022).

To address some of these challenges, some scholars suggest effective leadership, a strategic research agenda to obtain insights on the trip planning process and behavior of a visitor and, management protocol of tourism suppliers. Both tourists and stakeholders/suppliers of heritage tourism will need to understand, embrace and invest in new innovations so that value co-creation can be enhanced for all stakeholders in an equitable manner. To sum up, political support, investment and strategic leadership are required in building the four essential blocks of Metaverse:

1 technology-ready users (critical is willingness and readiness on the part of consumers in embracing Metaverse as a popular dais for connecting with tourism destinations and enterprises, seeking information and buying tourism products, etc.);
2 empowering platforms (virtual worlds);
3 the networking infrastructure (a critical building block "as low-speed Internet networks represent a major roadblock") and;
4 enabling devices (such as "MR/VR headsets, haptics and environment rendering devices").

(Buhalis et al., 2023, pp. 7–8)

It is critical to make the entire digital ecosystem affordable and sustainable in terms of mindfulness toward other users, onsite resources, the host community, and self and communal (refers to other users, social networks at place of residence and new bonds at the visited destination) well-being.

The Well-being and Transformation of the Host Community

This section focuses on ways in which individual and communal and ecological well-being can be achieved by pursuing a transformation pathways. As mentioned in the foregoing, research on the impact of heritage tourism on the well-being of host communities is meager. According to Brooks et al., the success of heritage tourism "depends on the inclusion of host communities in heritage tourism governance, decision-making processes, and access to resources and programs" (2023, p. 1). Recognizing the important association between heritage, tourism, and sustainable development, UNESCO launched the World Heritage and Sustainable Tourism Programme, which was espoused by the World Heritage Committee. It advocates a synergistic approach toward strategic planning between tourism and heritage management, to safeguard both cultural and natural resources and devise relevant sustainable tourism avenues (UNESCO World Heritage Center, 2021). According to Brooks et al., "by drawing on the cultural and historical capital of a community, heritage tourism can contribute to the flourishing of local communities and their positive sustainable development. However, when applied uncritically and without meaningful engagement with the needs of local stakeholder, heritage tourism can also elicit damaging effects on community health and wellbeing" (2023, pp. 2–3).

Brooks et al. (2023) examine the impacts of heritage tourism on host communities' ability to thrive and sustain its health and well-being over an extended period of time. The authors note that the most profound impact is the fear that they will be compelled to relocate from their ancestral land because of tourism or government-led conservation programs (Melubo & Lovelock, 2019; Stone, 2013). This threat has endangered their lifestyle and livelihood. Furthermore, it can result in far reaching social consequences in terms of compromising the viability of their cultural traditions and customs they practice on their land (Melubo & Lovelock, 2019; Stoddart et al., 2018; Stone, 2013; Su et al., 2016). It can also lead to disintegration of family structures and segregated communities. Similar concerns are pointed by Sawhney et al. (2024) in Chapter 7.

Most studies have examined the impact of heritage tourism on the visitors in terms of health and overall well-being (Bauer, 2017; Kunasegaran et al., 2019). Only recently, the health needs and well-being of the host community are being acknowledged as an inherent function of cultural heritage management and viable community development (Taçon & Baker, 2019). A plethora of studies surmise that the cultural institutions, such as museums, hold potential to enhance the health and well-being of host communities in way of stimulating calmness and enthusing positive physiological and affective changes and promoting contemplation besides providing reassurance for public health benefits (Ander et al., 2013; Orthel, 2022; Sofaer et al., 2021). Along similar lines, traditional medicine can play a profound role in enhancing the healthiness and well-being of communities who use it (Atal & Atal, 2024) besides promoting them to national and international markets to stimulate economic benefits (Riordan & Schofield, 2015). Based on their review of extensive literature, Brooks et al. note that very few studies have investigated factors that can contribute or deteriorate cultural health due to tourism and such studies have shared profound "negative and traumatising outcomes of tourism development, such as a rise of ethnoreligious conflict, loss of ancestral land, a dilution of cultural practices....... and a loss of cultural authenticity" (2023, p. 29). Socio-cultural dimensions of well-being, therefore, constitute a critical area of inquiry.

Just as paralleling community health and well-being with economic prosperity is dubious, it cannot be assumed that all health issues are limited to clinical care and disease, because health can also have cultural connotations (Napier et al., 2014). In fact, the UNESCO's report suggests that culture and heritage play a pivotal role in contributing to community health and well-being (UNESCO, 2019, 2020). This notion was earlier stressed by the by the World Heritage Convention (World Heritage Commission on Culture and Development, 1995). Reports from these organizations call for research focus on health and well-being from a cultural standpoint. Of prominence is the 1980s slogan "*nothing about us without us*" which emerged from a variety of health-related educational subjects such as public health, studies on disability and occupational therapy (Brooks et al., 2023). Therefore, it is important to seek insights on how well-being is understood by the host communities. Furthermore, "meaningful colonizing approaches to heritage tourism. More than half of the negative consequences of heritage tourism for health and well-being could be mitigated with policy guidance, contingent cultural protocols and anti-colonial methods" (Brooks et al., 2023, p. 30). It is also envisaged that "communities hold a range of knowledges and cultural assets that .... should be legally protected to deploy (or not) as a culturally-suitable foundation that steers the design of locally-governed tourism enterprises" (Brooks et al., 2023, p. 31); and to optimize benefits resulting from heritage tourism, presence and participation of host communities on topics associated with tourism, governance and, regulation of or access to cultural and natural resources/programs should be accorded precedence. Empowerment of local community groups is emphasized by Xie (2024) and Jolliffe et al. (2024) in Chapters 3 and 4.

Sustainable Marketing Doctrine of Heritage Tourism: Toward Transformed Consciousness

Recent studies argue that government intervention is needed to ensure that socially responsible policies are integrated in the vision and implemented through a range of consumer-centric initiatives of enterprises. Furthermore, it is posited that consumer consciousness and self-guided ethical consumption will not suffice to bring a transformative change in the tourism system. In other words, stakeholders need to collectively integrate and advocate ethical use of resources and their efforts should be regulated through political intervention. Partnerships and stakeholder interdependence are edifices to sustainable development, and "multi-stakeholder partnership networks can perform a key role in fostering social welfare with appropriate governance strategies in place" (Laczniak & Murphy, 2006, p. 167). The optimal aim is to develop a regulated transformational tourism system. Tactics and procedures of imparting knowledge to consumers and other marketing role players to attain an elevated sense of consciousness (eudaimonia) and developing mindfulness of the socio-ecological implications of both production/consumption behaviors have received limited attention in marketing literature, particularly in the tourism field. This chapter points out that both marketers and governments (as well as the civil society) have vested interests in "redirecting customer needs toward sustainable products, responsible consumption, eco-efficiency, eco-balance, and eco-sufficiency. Not only should businesses and marketers become more responsible for their activities, but they also need to be guided and directed through government policies and programs" (Sheth & Parvatiyar, 2021, p. 160). In other words, the entire heritage tourism system should be guided by the ethics theory to ensure socially responsible and sustainable production, marketing and consumption of fragile resources (Chhabra, 2012, 2021a; Timothy, 2021). This chapter suggests the path of "dharma," through strategic governance, toward eudaimonia (higher state of consciousness) to foster socially responsible sustainable practices which can lead to transformation of the whole heritage tourism system.

As mentioned in Chapter 1, Sheth and Parvatiyar (2021, 2022) point out that "dharma-related" righteousness needs to be practiced to produce desired outcomes in marketing. For instance, a theory of "dharma" (duties) and "karma" (actions) envisaged in a "less explored Eastern ethical system can provide a strong mandate for businesses to embrace socially responsible marketing" (Laczniak & Shultz, 2022, p. 558). Sheth and Parvatiyar take a corporate-level approach albeit this path is also relevant for the non-profit and public enterprises of the heritage tourism system. A brief overview of the path is offered in the next paragraph. Referring to the SDGs, Laczniak and Shultz speak that these seventeen goals, "while having an environmental tenor, go far beyond ecological concerns in their aspiration, and include working conditions, economic growth, the nature of responsible production and poverty level" (2022, p. 555). The fundamental tenets of socially responsible marketing (citizenship, stakeholder orientation, and social/ecological sustainability), adopted as the enduring righteousness of "dharma," may evolve with time but their rootedness in distributive justice will continue to inspire fruitful engagement of the social promise. Dharma refers to the ethical duties of marketing. Karma, however, postulates that every human action has a consequence and marketing should perform its duties in a manner that contributes to the society in terms of equity and well-being. Therefore, one pathway to deliver socially responsible marketing in heritage tourism is to embrace the notions of dharma and karma to attain the optimal eudaimonic outcome (Sheth & Parvatiyar, 2022). The components of SRM (socially responsible marketing) can be integrated into the sustainable marketing paradigm.

In the first chapter, I propose a guided sustainable marketing paradigm and in the second chapter, I present a transformative heritage tourism framework where I underline the significance of attaining eudaimonia (the optimal state of mind). Guided sustainable marketing agenda suggested in Chapter 1 is devised to support the conceptual path toward transformative heritage tourism which stems from Huang et al.'s (2022) needs-activities-well-being sequential path of transformation (please see Figure 2.1). Figure 12.1 illustrates how eudaimonia rooted in dharma and karma, guided by regulatory policies and appropriate governance, can support socially responsible sustainable marketing of heritage tourism (please see Figure 1.1) and the eventual transformation of the whole system. Figure 12.1 also emphasizes the role of supportive governance structures and political intervention to regulate the process. The next few paragraphs provide more insights into the "dharma," "karma," and the eudaimonia notions.

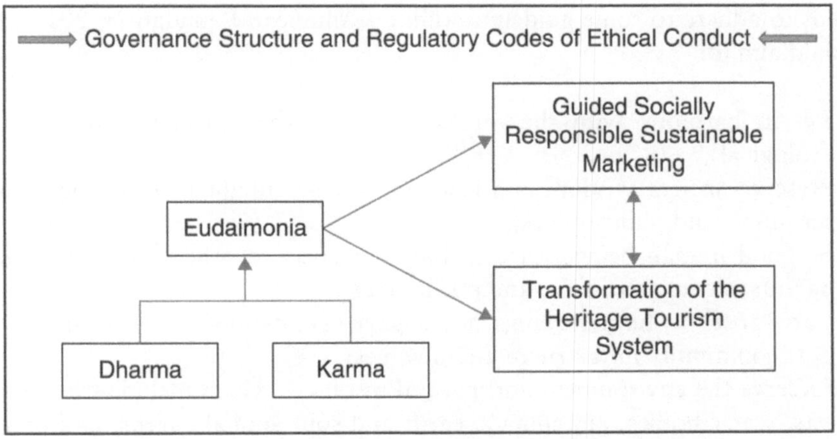

FIGURE 12.1 Transformative sustainable marketing doctrine

Eudaimonia

Quest for this state of mind has soared following the disenchantment triggered by capitalism and consumerism, the devastating pandemic and the desire to promote and nurture the well-being of the society (Chang & Durante, 2022; Ryan & Deci, 2001; Thompson & Kumar, 2022; Waterman, 2013). Sheth and Parvatiyar describe eudaimonia as "an ancient ethical concept, rooted in intellectual and moral virtues of wisdom, justice, and temperance" (2022, p. 591). It is being widely recognized that eudaimonia should be the optimal aim of marketing. It denotes quality of life and societal well-being by representing the higher order purpose that can serve to shape the activities/performance of the marketing personnel (Sirgy, 2021). An eudaimonic perspective will ensure that marketing contributes to sustained happiness and flourishing by rooting itself in the righteousness virtues ("dharma") to build a more "worthy world" and contributing toward a holistic life (Sheth & Parvatiyar, 2021).

Dharma

Sheth and Parvatiyar (2022) posit that the marketing's "dharma" (responsibility) is to pursue societal norms of righteous behavior. Dharma is defined as an obligation toward the "Universal Righteousness" from a moral perspective. The crux of dharma is grounded in the principle that human beings are "are indebted to the cosmic world and society (including parents, ancestors, teachers, community, and employers) who have made it possible to enjoy all the material and worldly benefits. As such, we are duty-bound and obligated to repay this debt" (Sheth & Parvatiyar, 2022, pp. 594, 595). Dharmic life needs to adhere to some guidelines (Hindu American Foundation, 2014). It should aim to:

1. Act in harmony with the forces sustaining the universe (cosmic and ecological);
2. Preserve ancient wisdom and teachings of the enlightened (in numerous scriptures and philosophies);
3. Be good human beings, care and show respect to others, including our parents, elders, ancestors, and our heritage;
4. Care for fellow human beings, and engage in community service, and support community infrastructure and values;
5. Preserve the environment and natural resources which include trees, forests, water bodies, air, climate, earth and soil, animals, birds, and other wild creatures.

(Sheth & Parvatiyar, 2022, p. 595)

To sum up the above, the aforementioned implies that sustainable marketing has a social obligation to harmonize with the universal realm and nurture all human beings, advocate community service, and protect the environment and its cultural and natural reserves. While "dharma" aims to inspire consciousness at the soul level, marketing strategies can karmically guide the goals and aspirations of societal well-being and socio-ecological sustainability (Laczniak & Schulz, 2022). Positive outcomes (or reduction of externalities) can be anticipated if marketing actions ("karma") are carried out mindfully based on the principles of "dharma" (Sheth & Parvatiyar, 2022). The proposed model, in Figure 12.1, emphasizes on harmony between the goals of different stakeholders and design/implementation of ethical codes of conduct (procedures) that can serve to guide decisions at the individual and group level to achieve equitable outcomes (distributive justice) for all. Support for procedural (acceptable codes of conduct) and distributive justice can foster trust in stakeholders and promote reciprocated relationships (Chhabra, 2023; Laczniak & Murphy, 2018; Lucas et al., 2016).

Summary

Several studies have made a dedicated effort to illustrate the manner in which marketing can play a key role in promoting sustainable communities, responsible corporate practices, and overall well-being of a broad spectrum of stakeholders of tourism. In a nutshell, sustainable marketing calls for accountable and pre-emptive business strategies to promote market demand toward sustainable behavior and purchase of sustainable products and services and foster the well-being of all stakeholders and the planet

Earth. Several studies argue that sustainable inclinations of the consumers alone cannot achieve meaningful sustainable outcomes without the support of the corporate sector and regulatory policies (Laczniak & Shultz, 2021). This chapter suggests a socially responsible integrated sustainable marketing doctrine for transformative heritage tourism by drawing on Sheth and Parvatiyar's (2022) unique eudaimonic approach. This approach is premised on the notion that the quest for overall wellness and flourishing is key to sustainable behavior. For enterprises to promote sustainable behavior of consumers, they have to first transform their internal culture and vision (Bolton, 2020; Chhabra, 2021b; Chhabra & Zhao, 2015; Huang & Rust, 2011). This chapter argues that transformative agenda is required to achieve eudaimonic centered socially responsible objectives to foreground the path toward successful sustainable initiatives and mindful guest and host behavior.

The core of responsible marketing should be guided by the collective moral values to generate synchronization at the cosmic and community level (Sheth & Parvatiyar, 2022). In congruence with the views of Laczniak and Shultz (2021), socially responsible practices require a civic/society-based rather than a personalized perspective. Specifically, Sheth and Parvatiyar (2022, p. 27) write that "marketing excellence should be viewed as a normative macro concept that achieves the subjective results of societal well-being." Underpinned on these views, a eudaimonic-centric integrated transformative sustainable marketing precept for heritage tourism can prove to be beneficiary for heritage institutions and agencies. Going forward, there is tremendous scope for the public and private sectors, the host communities and various custodians of local/indigenous culture and heritage to shape the consciousness of the entire heritage tourism system toward transformed sustainability and holistic well-being. Obstacles and challenges toward an elevated sense of global consciousness are not insurmountable. As illustrated in numerous chapters of this book, possibilities for change are more likely to come from regulatory policies and below, at a local scale. The dialogue initiated by the contributors of this book can be carried forward to build meaningful discourses on sustainability, sustainable marketing and transformative heritage tourism. I hope this book will be used as stepping stone for further dialogue and research.

References

Adie, B. A., & de Bernardi, C. (2023). 'Oh my god what is happening?': Historic second home communities and post-disaster nostalgia. *Journal of Heritage Tourism*, *18*(3), 337–348.

Albrecht, J. N., & Raymond, E. (2021). National destination pledges: Visitor management through emotional engagement, commitment and interpretation. *Tourism Management Perspectives*, *40*(4), 100894.

Allaberganov, A., & Catterall, P. (2023). Using social exchange theory to examine residents' responses to heritage tourism: Case studies of Samarqand and Bukhara in Uzbekistan. *Journal of Heritage Tourism*, *18*(6), 846–863.

Altschuler, W. (2024). Most surprising travel trends. Retrieved January 2024, from https://www.forbes.com/sites/wendyaltschuler/2024/02/21/the-hottest-travel-trends-that-will-surprise-you/?sh=6f5c9cd18194

Ander, E., Thomson, L., Noble, G., Lanceley, A., Menon, U., & Chatterjee, H. (2013). Heritage, health and well-being: Assessing the impact of a heritage focused intervention on health and well-being. *International Journal of Heritage Studies*, *19*(3), 229–242.

Atal, N., & Atal, C. (2024). Healing Potential of Herbs and Herbal Tourism. In D. Chhabra, N. Atal, & A. Maheshwari (Eds.), *Sustainable development and resilience of tourism: Wellbeing and quality of life perspectives*. Springer. In Press.

Basaraba, N. (2024). The rise of paranormal investigations as virtual dark tourism on YouTube. *Journal of Heritage Tourism*, *19*(2), 287–309.

Bauer, I. L. (2017). Improving global health–is tourism's role in poverty elimination perpetuating poverty, powerlessness and 'ill-being'? *Global Public Health*, *12*(1), 45–64.

Bolton, R. N. (2020). Commentary: Future directions of the service discipline. *Journal of Services Marketing*, *34*(3), 279–289.

Breakwell, G. M. (1986). *Coping with threatened identity*. Taylor & Francis.

Brewer, M. (1991). The social self: On being the same and different at the same time. *Personality and Social Psychology Bulletin*, *5*, 475–482.

Brooks, C., Waterton, E., Saul, H., & Renzaho, A. (2023). Exploring the relationships between heritage tourism, sustainable community development and host communities' health and wellbeing: A systematic review. *PLoS One*, *18*(3), e0282319.

Buhalis, D., Leung, D., & Lin, M. (2023). Metaverse as a disruptive technology revolutionising tourism management and marketing. *Tourism Management*, *97*, 104724.

Butler, G., Khoo-Lattimore, C., & Mura, P. (2014). Heritage tourism in Malaysia: Fostering a collective national identity in an ethnically diverse country. *Asia Pacific Journal of Tourism Research*, *19*(2), 199–218.

Chang, H., Zhang, L., & Xie, G. (2015). Message framing in green advertising: The effect of construal level and consumer environmental concern. *International Journal of Advertising*, *34*(1), 158–176.

Chang, Y., & Durante, K. M. (2022). Why consumers have everything but happiness: An evolutionary mismatch perspective. *Current Opinion in Psychology*, *46*, 101347.

Chen, S., Boucher, H. C., & Tapias, M. P. (2006). The relational self revealed: Integrative conceptualization and implications for interpersonal life. *Psychological Bulletin*, *132*(2), 151–179.

Chen, S., Lee-Chai, A. Y., & Bargh, J. A. (2001). Relationship orientation as a moderator of the effects of social power. *Journal of Personality and Social Psychology*, *80*, 173–187.

Chen, Z., & Wan, P. Y. K. (2023). Interdependence between Community participation and sustainable livelihood development: Community perspectives on heritage conservation in Macao. *Journal of Heritage Tourism*, *18*(1), 1–17.

Chhabra, D. (2012). A present-centered dissonant heritage management model. *Annals of Tourism Research*, *39*(3), 1701–1705.

Chhabra, D. (2017). Heritage branding of India: A Gandhi tourism perspective. *Annals of Tourism Research*, *62*, 110–112.

Chhabra, D. (2021a). Transformative perspectives of tourism: Dialogical perceptiveness. *Journal of Travel & Tourism Marketing*, *38*(8), 759–768.

Chhabra, D. (2021b). *Resilience, authenticity and digital heritage tourism*. Routledge.

Chhabra, D. (2023). Modelling heritage justice for under-represented communities. In *Routledge handbook of trends and issues in global tourism supply and demand* (pp. 148–162). Routledge.

Chhabra, D., & Zhao, S. (2015). Present-centered dialogue with heritage representations. *Annals of Tourism Research*, *55*, 94–109. (One cite needed)

Chi, O. H., Denton, G., & Gursoy, D. (2021). Interactive effects of message framing and information content on carbon offsetting behaviors. *Tourism Management*, *83*(2), 104244.

Christou, P. A., Timothy, D. J., & Pericleous, K. (2024). Reflections on tourists' experiential engagement with Scotland's whisky heritage. *Journal of Heritage Tourism*, *19*(2), 201–220.

Cordina, R., Gannon, M., Taheri, B., Okumus, F., & Lochrie, S. (2022). Committed to conservation: Tourism in developed and developing contexts. *International Journal of Tourism Research*, *24*(2), 323–336.

del Mármol, C., & Santamarina, B. (2024). 'We gave the city its image and put it on the map.': Intangible cultural heritage and city branding in Buenos Aires and Valencia. *Journal of Heritage Tourism*, *19*(2), 243–262.

Dragin, A., Mijatov Ladičorbić, M., Stojanović, V., Lazić, L., Komnenić, B., Damnjanović, J., …, & Hansen, B. (2023). Evaluating resident attitudes on German industrial heritage and tourism using the sustainable tourism attitude scale in Vojvodina, Serbia. *Journal of Heritage Tourism*, *18*(6), 864–884.

Duda, T. (2023). The challenges and new perspectives on re-interpreting the transboundary heritage for tourism, an example of the Griffins' Route, South Baltic area. *Journal of Heritage Tourism*, *18*(1), 84–100.

Earl, A., & Hall, C. M. (2023). Nostalgia and tourism. *Journal of Heritage Tourism*, *18*(3), 307–317.

Fagence, M. (2024). The challenge for transformative tourism of difficult and dissonant heritage. In D. Chhabra (Ed.), *Sustainable marketing of transformative heritage tourism*. Routledge. In Press.

Farrelly, F., Kock, F., & Josiassen, A. (2019). Cultural heritage authenticity: A producer view. *Annals of Tourism Research*, *79*, 102770.

Fagence, M. (2024). The Challenge for Transformative Tourism of Diffcult and Dissonant Heritage. In D. Chhabra (Ed.), (Ed.), *Sustainable marketing of transformative heritage tourism*. Routledge. In Press.

Geertz, C. (2017). Ritual and social change: A Javanese example. In P. Stewart, & A. Strathern (Eds.), *Ritual* (pp. 549–576). Routledge.

Grazzini, L., Rodrigo, P., Aiello, G., & Viglia, G. (2018). Loss or gain? The role of message framing in hotel guests' recycling behaviour. *Journal of Sustainable Tourism*, *26*(11), 1944–1966.

Hashimoto, A., & Telfer, D. J. (2024). Bending history into Heritage at hidden Christian sites in Nagasaki, Japan: UNESCO World Heritage site designation and tourism. *Journal of Heritage Tourism*, *19*(2), 221–242.

He, L. (2024). Transformative tea tourism in Northern Thailand. In D. Chhabra (Ed.), *Sustainable marketing of transformative heritage tourism*. Routledge.

He, X., Cheng, J., Swanson, S. R., Su, L., & Hu, D. (2022). The effect of destination employee service quality on tourist environmentally responsible behavior: A moderated mediation model incorporating environmental commitment, destination social responsibility and motive attributions. *Tourism Management*, *90*(3), 104470.

Helgeson, V. S. (1994). Relation of agency and communion to well-being: Evidence and potential explanations. *Psychological Bulletin*, *116*(3), 412–428.

Hindu American Foundation (2014). What is dharma? https://www.hinduamerican.org/wp-content/uploads/2019/12/Dharma2.0_1.pdf

Huang, M. H., & Rust, R. T. (2011). Sustainability and consumption. *Journal of the Academy of Marketing Science*, *39*, 40–54.

Hunter, W. C. (2011). Rukai indigenous tourism: Representations, cultural identity and Q method. *Tourism Management*, *32*(2), 335–348.

Hussein, A. S., & Hapsari, R. (2021). Heritage experiential quality and behavioural intention: Lessons from Indonesian heritage hotel consumers. *Journal of Heritage Tourism*, *16*(3), 317–336.

Jaafar, M., Noor, S. M., & Rasoolimanesh, M. (2015). Perception of young local residents toward sustainable conservation programmes: A case study of the Lenggong World Cultural Heritage Site. *Tourism Management*, *48*, 154–163.

Jolliffe, L., Piboonrungroj, P., & Chaikor, J. (2024). Transformative tea tourism in Northern Thailand. In D. Chhabra (Ed.), *Sustainable marketing of transformative heritage tourism*. Routledge. In Press.

Keyes, C. L., Shmotkin, D., & Ryff, C. D. (2002). Optimizing well-being: The empirical encounter of two traditions. *Journal of Personality and Social Psychology*, *82*(6), 1007.

Kifer, Y., Heller, D., Perunovic, W., & Galinsky, A. D. (2013). The good life of the powerful: The experience of power and authenticity enhances subjective well-being. *Psychological Science*, *24*(3), 280–288.

Kunasegaran, M., Rasoolimanesh, S. M., & Khan, S. K. (2019). Experiences of international tourists with healthy sig-nature foods: A case study in Malacca. *British Food Journal*. https://doi.org/10.1108/BFJ-08-2018-0567

Kwan, V., Bond, M., & Singelis, T. (1997). Pancultural explanations for life satisfaction: Adding relationship harmony to self-esteem. *Journal of Personality and Social Psychology*, *73*(5), 1038–1051.

Laczniak, G. R., & Murphy, P. E. (2006). Normative perspectives for ethical and socially responsible marketing. *Journal of Macromarketing*, *26*(2), 154–177.

Laczniak, Gene R., & Patrick E. Murphy (2018). The role of normative marking ethics. *Journal of Business Research*, *95*, 401–407.

Laczniak, G., & Shultz, C. (2021). Toward a doctrine of socially responsible marketing (SRM): A macro and normative-ethical perspective. *Journal of Macromarketing*, *41*(2), 201–231.

Laczniak, G., & Shultz, C. (2022). Socially responsible marketing (SRM): Broader, deeper, and ever greater. *Journal of Macromarketing*, *42*(4), 554–559.

Li, Y., Wang, X., & Chen, P. (2019). Development and testing of heritage responsibility scale: Based on perspectives of tourists and residents. *Tourism Tribune*, *34*(10), 60–75.

Lucas, T., Kamble, S. V., Wu, M. S., Zhdanova, L., & Wendorf, C. A. (2016). Distributive and procedural justice for self and others: Measurement invariance and links to life satisfaction in four cultures. *Journal of Cross-Cultural Psychology*, *47*(2), 234–248.

Lund, N. F., Holst-Beck, P., Pine, B. J. II, & Leask, A. (2022). Hamlet Live: The 7 I's experiential strategy framework for heritage visitor attractions. *Journal of Heritage Tourism*, *17*(2), 125–141.

Madgin, R., & Lesh, J. (2021). *People-centred methodologies for heritage conservation*. Routledge.

Melubo, K., & Lovelock, B. (2019). Living inside a UNESCO World Heritage Site: The perspective of the Maasai community in Tanzania. *Tourism Planning & Development*, *16*(2), 197–216.

Moon, H., & Han, H. (2019). Tourist experience quality and loyalty to an island destination: The moderating impact of destination image. *Journal of Travel & Tourism Marketing*, *36*(1), 43–59.

Musikanski, L., Rogers, P., Smith, S., Koldowski, J., & Iriarte, L. (2019). Planet happiness: A proposition to address overtourism and guide responsible tourism, happiness, well-being and sustainability in world heritage sites and beyond. *International Journal of Community Well-Being*, *2*(3), 359–371.

Napier, A. D., Ancarno, C., Butler, B., Calabrese, J., Chater, A., & Chatterjee, H., et al. (2014). Culture and health. *The Lancet*, *384*(9954), 1607–1639.

Neff, K. D., & Harter, S. (2002). The role of power and authenticity in relationship styles emphasizing autonomy, connectedness, or mutuality among adult couples. *Journal of Social and Personal Relationships*, *19*(6), 835–857.

Nguyen, K. T., Murphy, L., Chen, T., & Pearce, P. L. (2024). Let's listen: The voices of ethnic villagers in identifying host-tourist interaction issues in the Central Highlands, Vietnam. *Journal of Heritage Tourism*, *19*(2), 263–286.

Nikolaou, P. (2024). Museums and the post-digital: Revisiting challenges in the digital transformation of museums. *Heritage*, 7(3), 1784–1800.

Orthel, B. (2022). Linking public health and heritage work. *International Journal of Heritage Studies*, *28*(1), 44–58.

Paulhus, D. L., & Martin, C. L. (1988). Functional flexibility: A new conception of interpersonal flexibility. *Journal of Personality and Social Psychology*, *55*(1), 88.

Peirson, E. (2023). Remembering Folkestone: Collective nostalgia and sense of place in the English seaside resort. *Journal of Heritage Tourism*, *18*(3), 318–336.

Peng, K., & Nisbett, R. E. (1999). Culture, dialectics, and reasoning about contradiction. *American Psychologist*, *54*(9), 741.

Pettersson, R., & Müller, D. K. (2023). Museums portraying indigenous heritage: The case of Sámi museums in Sweden. *Journal of Heritage Tourism*, *18*(2), 184–201.

Pew Research Center. (2022). *Metaverse in 2040*. Pew Research Center.

Pintel, T. & Chhabra, D. (2024). Social tagging museum collections: Public engagement or an Abrogation Act. In D. Chhabra (Ed.), *Sustainable marketing of transformative heritage tourism*. Routledge. In Press.

Pradhan, S., Dyson, L. E., & Lama, S. (2022). The nexus between cultural tourism and social entrepreneurship: A pathway to sustainable community development in Nepal. *Journal of Heritage Tourism*, *17*(6), 615–630.

Pratt, M. (2022). Metaverse pros and cons: Top benefits and challenges. Retrieved from https://www.techtarget.com/searchcio/tip/Metaverse-pros-and-cons-Top-benefits-and-challenges

Prayag, G., & Del Chiappa, G. (2023). Nostalgic feelings: Motivation, positive and negative emotions, and authenticity at heritage sites. *Journal of Heritage Tourism*, *18*(3), 349–364.

Qin, H. X., Wang, Y., & Hui, P. (2022). Identity, crimes, and law enforcement in the metaverse. *arXiv preprint arXiv:2210.06134*.

Riordan, A., & Schofield, J. (2015). Beyond biomedicine: Traditional medicine as cultural heritage. *International Journal of Heritage Studies*, *21*(3), 280–299.

Roque, M. I., Campos, A. C., Almeida, S., & Pasandideh, S. (2024). Transforming museum management through ICT adoption: An analysis of the Portuguese context during the COVID-19 pandemic. *Journal of Heritage Tourism*, 1–18.

Ryan, R. M., & Deci, E. L. (2001). On happiness and human potentials: A review of research on hedonic and eudaimonic well-being. *Annual Review of Psychology*, *52*(1), 141–166.

Ryu, K., Lehto, X. Y., Gordon, S. E., & Fu, X. (2019). Effect of a brand story structure on narrative transportation and perceived brand image of luxury hotels. *Tourism Management*, *71*(2), 348–363.

Santa, E. D., & Tiatco, S. A. (2019). Tourism, heritage and cultural performance: Developing a modality of heritage tourism. *Tourism Management Perspectives*, *31*(3), 301–309.

Sawhney, R., Alka Maheshwari, A., & Gnanasambandan, S. (2024). Tribal tourism: A regulatory based approach towards empowerment of tribes. In D. Chhabra (Ed.), *Sustainable marketing of transformative heritage tourism*, Routledge, In Press.

Schmid, P. F. (2001). Authenticity: The person as his or her own author. Dialogical and ethical perspectives on therapy as an encounter relationship. And beyond. Rogers' therapeutic conditions: *Evolution, Theory and Practice*, *1*, 213–228.

Schmid, P. F. (2005). Authenticity and alienation: Towards an understanding of the person beyond the categories of order and disorder. *Person-Centred Psychopathology*, *3*, 75–90.
Sheldon, K., Ryan, R. M., Rawsthorne, L. J., & Ilardi, B. (1997). Trait self and true self: Crossrole variation in the Big-Five personality traits and its relations with psychological authenticity and subjective well-being. *Journal of Personality and Social Psychology*, *73*(6), 1380–1393.
Sheth, J. N., & Parvatiyar, A. (2021). Sustainable marketing: Market-driving, not market-driven. *Journal of Macromarketing*, *41*(1), 150–165.
Sheth, J. N., & Parvatiyar, A. (2022). Socially responsible marketing: Toward aligning dharma (duties), karma (actions), and eudaimonia (well-being). *Journal of Macromarketing*, *42*(4), 590–602.
Sirgy, M. J. (2021). Macromarketing metrics of consumer wellbeing: An update. *Journal of Macromarketing*, *41*(1), 124–131.
Sofaer, J., Davenport, B., Sørensen, M. L. S., Gallou, E., & Uzzell, D. (2021). Heritage sites, value and wellbeing: Learning from the COVID-19 pandemic in England. *International Journal of Heritage Studies*, *27*(11), 1117–1132.
Stoddart, M., Catano, G., & Ramos, H. (2018). Navigating tourism development in emerging destinations in Atlantic Canada: Local benefits, extra-local challenges. *Journal of Rural and Community Development*, *13*(2), 57–75.
Stone, M. (2013). Protected Areas, Tourism and Rural Community Livelihoods in Botswana [PhD Thesis]: Arizona State University.
Su, M., Wall, G., & Xu, K. (2016). Tourism-induced livelihood changes at Mount Sanqingshan World Heritage Site, China. *Environmental Management*, *57*(5), 1024–1040.
Szromek, A. R., & Butler, R. W. (2024). A re-positioning of post-industrial heritage in upper Silesia, Poland, into an integrated thematic tourist route. *Journal of Heritage Tourism*, *19*(2), 310–330.
Taçon, P. S. C., & Baker, S. (2019). New and emerging challenges to heritage and well-being: A critical review. *Heritage*, *2*(2), 1300–1315.
Tan, S., Tan, S., Kok, Y., & Choon, S. (2018). Sense of place and sustainability of intangible cultural heritage – The case of George Town and Melaka. *Tourism Management*, *67*, 376–387.
Taylor, M. (2024). Travel trends that will define 2024. Retrieved January 2024, from https://www.travelpulse.com/news/features/the-travel-trends-that-will-define-2024
Thirumaran, K., Mohammadi, Z., Azzali, S., Eijdenberg, E. L., & Donough-Tan, G. (2023). Transformed landscapes, tourist sentiments: The place making narrative of a luxury heritage hotel in Singapore. *Journal of Heritage Tourism*, *18*(2), 243–264.
Thompson, C. J., & Kumar, A. (2022). Analyzing the cultural contradictions of authenticity: Theoretical and managerial insights from the market logic of conscious capitalism. *Journal of Marketing*, *86*(5), 21–41.
Thouki, A. (2024). How authenticity is negotiated and experienced by clergy and tourists at religious World Heritage sites: Taking a closer look into the 'preserve as found' strategy. *Journal of Heritage Tourism*, 1–25.
Tiberghien, G., Bremner, H., & Milne, S. (2020). Authenticity and disorientation in the tourism experience. *Journal of Outdoor Recreation and Tourism*, *30*, 100283.
Timothy, D. J. (2021). *Cultural heritage and tourism: An introduction* (2nd ed.). Channel View Publications.
Trupp, A., Shah, C., & Hitchcock, M. (2023). Globalisation, crafts, and tourism microentrepreneurship in the South Pacific: Economic and sociocultural dimensions. *Journal of Heritage Tourism*, *18*(6), 733–755.
Tuan, Y.-F. (1974). *Topophilia*. Prentice-Hall.
Tuan, Y. F. (1977). *Space and place: The perspective of experience*. U of Minnesota Press.

UNESCO (2020). *Operational directives for the implementation of the convention for the safeguarding of the intangible cultural heritage*. UNESCO.

UNESCO World Heritage Centre (2019). *Operational guidelines for the implementation of the world heritage convention*. UNESCO.

UNESCO World Heritage Centre (2021). Sustainable tourism: UNESCO World Heritage and Sustainable Tourism Programme [cited 2024 March 10]. Available from https://whc.unesco.org/en/tourism/

Viana-Lora, A. (2024). Planning heritage tourism from a sustainable marketing perspective. In D. Chhabra (Ed.), *Sustainable marketing of transformative heritage tourism*. Routledge. In Press.

Vidal-Matzanke, A., Esteve-Ibáñez, H., & Vidal-González, P. (2023). Can cultural heritage be a successful strategy as a tourist experience? The case of tourist establishments in the mountains of Valencia, Spain. *Journal of Heritage Tourism*, *18*(2), 265–282.

Volchek, K., & Brysch, A. (2023, January). Metaverse and tourism: From a new niche to a transformation. In ENTER22 e-tourism conference (pp. 300–311). Springer Nature Switzerland.

Wali, N., Akombi, B. J., James, P., Waterton, E., Saul, H., Yuol, A., & Renzaho, A. (2019). The impact of heritage tourism on sustainable community development, health and wellbeing: A systematic review protocol. *Social Science Protocols*, *2*, 1–10.

Wang, Y. N. (2016). Balanced authenticity predicts optimal well-being: Theoretical conceptualization and empirical development of the authenticity in relationships scale. *Personality and Individual Differences*, *94*, 316–323.

Wang, M. Y., Li, Y. Q., Ruan, W. Q., Zhang, S. N., & Li, R. (2024). Influencing factors and formation process of cultural inheritance-based innovation at heritage tourism destinations. *Tourism Management*, *100*, 104799.

Wang, M. Y., Li, Y. Q., Ruan, W. Q., Zhang, S. N., Li, R., & Zhang, K. F. (2024). To understand or to touch? Evoking tourists' cultural preservation commitment through heritage tourism interpretation. *Journal of Sustainable Tourism*, 1–20.

Wang, R., & Quek, F. (2010, January). Touch & talk: contextualizing remote touch for affective interaction. In *Proceedings of the fourth international conference on Tangible, embedded, and embodied interaction* (pp. 13–20).

Wang, Q., & Witte, A. (2023). Hokkien Chinese diaspora visitors' image construction of their ancestral hometown: The role of the tourist gaze. *Journal of Heritage Tourism*, *18*(6), 768–784.

Waterman, A. S. (2013). The humanistic psychology–Positive psychology divide: Contrasts in philosophical foundations. *American Psychologist*, *68*(3), 124–133.

Wood, A., Linley, P., Maltby, J., Baliousis, M., & Joseph, S. (2008). The authentic personality: A theoretical and empirical conceptualization and the development of the authenticity scale. *Journal of Counseling Psychology*, *55*(3), 385–399.

World Commission on Culture and Development (1995). *Our creative diversity: Report of the world commission on culture and development*. UNESCO.

Wu, G. M., Chen, S. R., & Xu, Y. H. (2023). Generativity and inheritance: Understanding Generation Z's intention to participate in cultural heritage tourism. *Journal of Heritage Tourism*, *18*(4), 465–482.

Xie, P. (2024). Heritage tourism as transformation in staycation era. In D. Chhabra (Ed.), *Sustainable marketing of transformative heritage tourism*. Routledge. In Press.

Yi, X., Fu, X., Lin, B., & Sun, J. (2024). Authenticity, identity, self-improvement, and responsibility at heritage sites: The local residents' perspective. *Tourism Management*, *102*, 104875.

Yilmaz-dogan, O. and Ece, C. (2024). Rebranding or Retro-branding of Heritage Destinations In D. Chhabra (Ed.), *Sustainable marketing of transformative heritage tourism*. Routledge. In Press.

Yun, S. (2010). Cultural consciousness cultural confidence cultural self-improvement—Reflections on the prosperity and development of socialist culture with Chinese characteristics [J]. *Red Flag Manuscript*, (184), (16), 1–4.

Zhang, C. (2014). Heritage responsibility: Conception, characteristics and research issues. *Tourism Tribune/Lvyou Xuekan*, *29*(11), 45–51.

Zhang, M., Zhang, G., Gursoy, D., & Fu, X. (2018). Message framing and regulatory focus effects on destination image formation. *Tourism Management*, *69*(6), 397–407.

Zhou, Q., Zhang, J., Zhang, H., & Ma, J. (2015). A structural model of host authenticity. *Annals of Tourism Research*, *55*, 28–45.

INDEX

Note: Page numbers in **bold** and *italics* refer to tables and figures, respectively.